Lucy:
Growing Up Human

A Chimpanzee Daughter
in a Psychotherapist's Family

by
Maurice K. Temerlin

SCIENCE AND BEHAVIOR BOOKS, INC.   •   PALO ALTO, CALIFORNIA

To: *Julie,* My Jewish Mother;

Lucy, My Chimpanzee Daughter;

Jane, Lucy's Mother; and

Steve, Lucy's Brother.

TABLE OF CONTENTS

Preface   ix

Acknowledgements   xvii

Introduction   xxi

CHAPTER ONE
Lucy Comes Home   1

CHAPTER TWO
Chimpanzee Daughters and Jewish Mothers   31

CHAPTER THREE
What Will the Neighbors Think?   46

CHAPTER FOUR
Picnic   67

CHAPTER FIVE
Games Lucy Plays   85

CHAPTER SIX
Creative Masturbation   100

CHAPTER SEVEN
Can Animals Talk?   113

CHAPTER EIGHT
On Incest and Oedipus   126

CHAPTER NINE
Keeping It in the Family   144

CHAPTER TEN
Lucy As Co-Therapist   162

CHAPTER ELEVEN
Aggression and Hostility   174

CHAPTER TWELVE
Oedipus-Schmedipus   197

CHAPTER THIRTEEN
Tomorrow   205

Bibliography   213

PREFACE

The Kellogs were the first psychologists to study man's closest primate relative in the way we did. They raised a young chimpanzee named Gua with their son, Donald. They compared their growth rates, but discontinued the experiment after about a year.

Almost 25 years ago another family of psychologists, the Hayes, raised a female named Viki as their child for about six years. Unfortunately, Viki died before she reached sexual maturity. Thus, prior to our raising Lucy, no one to my knowledge has successfully raised a chimpanzee from early infancy to sexual maturity as a human child in a human home, isolated from her own species. Lucy thus is unique—if for no other reason. However, she is also unique for many other reasons. Her experience has had human reference points exclusively, so a study of her sexual and social behavior in adulthood can tell us a great deal when compared with the behavior of adult chimpanzees in nature. Since Lucy was raised as a human being, and since with humans sexuality is highly influenced by individual experience, leading to many different actions, I thought her sexuality would be directed toward humans. When she matured, Lucy's sexual preferences were directed toward humans—but in particular ways and toward particular humans I never would have predicted.

Unlike Gua and Viki, Lucy had a psychotherapist for a father. Psychotherapists make observations and draw inferences in ways very different from the thought processes of academic psychologists or other behavioral scientists. They usually have different values, epistemological criteria, and are interested in different subject matter; internal processes such as early experiences and human feelings, not just overt behavior, for example. This is one reason for the clinical-experimental or scientific-humanistic split which has characterized the history of psychology. Anyway, being a therapist I was primarily interested in how personality develops, grows, and changes—Lucy's and my own—with respect to these basic dimensions of human existence: sex; aggression and hostility; affection and love; attitudes toward the self and parents; and the control and integrative processes by which such inner psychological events are translated into overt action.

I was teaching and practicing psychotherapy while Lucy was growing up, and I was in psychotherapy as a patient several times during this period. Often I was growing as a person myself; more often I was stagnating. In either case, I tried to stay aware of what was going on in myself and in Lucy. In this book, I often write about my feelings and thoughts and therapy, perhaps as much as I write about Lucy. It would be impossible for me to do otherwise. I have never believed in the disembodied intellect, observing, correlating, and concluding with pure objectivity. Human values, feelings and emotions influence perceptions, as does the situation in which observations are made. And with living organisms as highly evolved and complicated as man and chimpanzee, the very act of observing (with whatever feeling) may change the behavior being observed. Many times this psychological Heisenberg principle has been frustrating to me, for example, when I was repeatedly unable to take pictures of Lucy masturbating with a Montgomery Ward vacuum cleaner.

Our lives have been organized around Lucy for almost ten years. We were constantly preoccupied with what she was doing, how she felt, what she might be thinking, her health and the many problems of child-rearing and protecting. One of us was always with Lucy. Jane and I never took a vacation together while Lucy was growing up. It is not as easy to get chimpanzee-sitters as it is to find baby-sitters. In spite of the social isolation and restrictions raising Lucy imposed upon us, some of the greatest joys of my life came from my years with Lucy. Lucy gave me many new insights, not only into herself or chimpanzees in general, but knowledge of myself as a human being and of the growth process, both in and out of psychotherapy.

As I watch her sitting on a sofa in our living room, nursing a gin and tonic while looking at the pictures in a magazine, and I observe myself doing the same thing, I can feel a kinship to the common ancestor we shared millions of years ago. Or when we are eating with her at the table: though she uses human silverware she eats so naturally, chewing with her mouth open and making sounds of such obvious delight, that I wonder if this is the way I would have been had I not been "civilized" within the anhedonic Judaic-Christian tradition. Looking into her clear brown eyes as she sits in my lap, her arms entwined about me, I often have the feeling that the deeper I look into Lucy the more I may see of my own basic nature. She helped me become more understanding and tolerant of myself and other people, for she takes herself and the humans she likes for granted, rejecting neither her own animal nature or her acquired humanness.

The original plan of this book was that it would be called *One Chimpanzee's Family* and divided into three sections with Jane, Steve, and myself each writing one. I was to describe what I as a therapist felt, thought, and observed as my daughter grew from a tiny, cuddly infant to the ninety pound full-grown humanized chimp she is today. Jane was to describe her experiences mothering Lucy, many aspects of which are completely unique. For example, it was clear throughout our ten years with Lucy that Jane felt about her as though she were her own offspring, the natural product of her body, rather than an adopted child from another species. All Jane's capacities for motherhood were mobilized and both her overt behavior toward Lucy and her inner experience demonstrate clearly that the act of parturition is in no way a prerequisite for the emergence of maternal behavior in the human female.

Steve also had a unique perspective from which to write about Lucy. He is the only human child to have spent ten years being raised in a human home with a chimpanzee as a sibling. In some ways Steve experienced the *Tarzan of the Apes* myth, living it out in reality, as in the Edgar Rice Burroughs' novels. He many times had the experience of going into the woods with a chimpanzee sister, and playing together as close to nature as a human being in our society can get. He taught Lucy to climb her first tree and to this day he treats Lucy as though she were a most treasured part of himself. He simply cannot keep his hands off of her, caressing her, tickling her, pushing her, nudging her, stroking her, so that she will love and play and wrestle with him. He now is twenty-one years old.

At the same time, writing his third of the book was acutely painful to him, and after a preliminary start he decided not to continue. With great reluctance I accepted Steve's decision. I understood it though I did not like it, and there was nothing I could do about it. There were times when they were growing up when I loved Lucy differently, if not more, than I did Steve. I can remember, for example, making such coarse and thoughtless remarks to Jane within Steve's hearing as, ". . . in the next reincarnation I would rather raise chimpanzees than human children, because they do not hold grudges or attempt to make their parents feel guilty when their needs are not met."

Further, and perhaps more important psychologically, the reactions of a father to his son reflect his attitudes toward himself. There were times during that ten years when I did not like myself, and at those times I rejected Steve in favor of Lucy, for I saw more of myself in him. I suspect that the pain which prevented him from writing his share has to do with a reexamination of his early relationship with me. Later, after the book was completed and a contract signed with the publisher, Steve told me there was another reason, one I should have realized but did not. His self-concept had been confused and chaotic. Often, he said, he could not tell the difference between himself and Lucy. External differences were obvious, of course; but they had been so close emotionally he had experienced a confusion of ego boundaries. There were times when Steve, with his eleven-year-old mind, felt that Lucy was human but that he was not. Though Steve has grown into a man I now treasure as a friend and, were I a child, would be proud to have as a father, there is a shyness about him which came from those ten years during which he felt most comfortable with "humans" such as Lucy.

Jane, too, declined to write her share. With Lucy and Steve grown, Jane has gone back to school. She is deeply involved in the women's movement and a career, and she did not want to collaborate on a book at that time. She said that she preferred to write her own book later, and she now is in the process of doing so.

The way I came to write this book is interesting, and for me it proves Jane is correct about the importance of timing. When we first adopted Lucy we did not intend to write a book—just to live, learn, and enjoy. But when the right time came the book just flowed out of me and I wrote the first draft in four months. It happened this way.

For twenty years I was a Professor of Psychology at a large midwestern state university noted for the excellence of its football team. For a five-year period I was chairman of its Psychology Department. The University throughout most of those years was a nice, quiet, easy-going place in which a faculty member could by and large do as he pleased if he had tenure and was not a communist, homosexual, or anti-football. It therefore was a comfortable place in which to teach and practice humanistic psychotherapy, which I liked far more than administration or research, even though I wrote articles when the impulse to do so struck me. But as everything changes, so did the University. The new administration emphasized research, wishing to change the image of the university. Like all enforced virtue or legislated morality, the substance got lost in favor of form or appearance. I began to feel they did not care about science as a process of discovering knowledge. Rather, they seemed to me interested only in the prestige and rituals of science. They seemed interested primarily in the number of publications, for example. Others seemed obsessed with proving what they already thought in the sense of "testing" hypotheses. The ones I knew usually "found" what they hypothesized because of experimenter-bias, pressures to publish, or the "demand characteristics" of the experiment; that is, the subject's capacity to perceive the experimental situation, try to read the experimenter's mind, and then to con him in some way. Though my classes attracted hundreds, who were charmed and enthralled by Lucy stories, this did not add to my credit. I was passed over for raises several times and before I resigned, the chariman of the department told me raises would return and he could even reduce my teaching load if I would start publishing again. He implied directly that it was *publication*, not teaching or research that counted with the new administration. Indeed, pressures to publish were so great, a very bad joke could produce hysterical laughter among the nonpublishing faculty:

"Did you hear that Christ returned and the Deans denied him tenure?"

"No, why?"

"They said he was a great teacher, but what did he ever publish?"

In my own field of humanistic psychotherapy, the psychology department was a sterile place anyway and twenty years in a comfortable rut is enough. So I resigned to enter full-time private practice.

I joined two old friends, Dick Sternlof and Ellen Oakes, and one new one, Nelda Ferguson. They had just built Timberridge, a school for children with learning and emotional difficulties. However, their building was a forty-five minute drive from my home and we could not move because of Lucy. So to practice psychotherapy I drove forty-five minutes in the morning and, after seeing eight to ten people each day, made the same drive home in the evening. I began to wonder how to amuse myself during the tedious drive. So I bought a Sony cassette recorder and dictated the first draft of the book. It came quite naturally. I sometimes could start dictating as I drove out of my driveway and the next thing that I was aware of was parking at my office, having dictated twenty-five or thirty pages. Sometimes these pages would be good enough that only minimal rewriting was required. More often they required considerable rewriting, but the content was there and it flowed out of me as though I were free associating in psychotherapy.

I hope that reading the book is as relaxed and natural a process for you as was the dictation of the first draft for me.

ACKNOWLEDGEMENTS

I am most grateful to Lucy, a unique being. Lucy is the only chimpanzee, to my knowledge, who has lived from birth through sexual maturity as a member of a human family. For her capacity to maintain her integrity as a chimpanzee while adapting to the joys and sorrows of the human condition, I am eternally grateful.

My wife, Jane, made the book possible. Without her exquisitely sensitive mothering of Lucy there would have been no material for a book. I am grateful to my son, Steve, who shared with me his early feelings toward his chimpanzee sister.

I appreciate also the help that I received from my brother, Liener Temerlin, of Dallas, Texas. Liener read the first and the last draft of the manuscript and made many helpful suggestions.

My friends at Timberridge, Dick Sternlof, Ellen Oakes, and Nelda Ferguson gave me constant support; they, too, read several chapters and made valuable suggestions. Linda Beverly gave me encouragement when I most needed it, during the first month I worked on the book. Mava Faulkner, Tommie Cochran, and Anna Burtnette created a pleasant and cordial atmosphere in which to work.

George I. and Judith Brown of Santa Barbara, California, gave me help in a variety of forms. Their workshop at Esalen helped crystalize the idea for the book and released the energy to write it. They each read several chapters and encouraged me to continue writing. George also gave me information about publishing.

I am enormously grateful to Drs. Roger Fouts and Sue Savage. Roger Fouts provided the gems from his sign language conversations with Lucy which are included in the chapter, "Can Animals Talk?" Furthermore, Roger was a warm and loving companion for Lucy during her language lessons. Sue Savage generously made available her observations of Lucy, including invaluable comparisons between Lucy and colony-raised chimps, which she painstakingly made during hundreds of hours of observation. Sue also read the manuscript, made helpful suggestions, and gave me many wonderful pictures of Lucy.

I want to express my gratitude to my parents. From my multilingual father I acquired skill with words, and from my mother I acquired whatever sensitivity and intelligence I may have. It is in this context that I want to honor a special obligation to Alan Jacobs of Newport Beach, California. Mr. Jacobs carefully read the entire manuscript and made many valuable suggestions. But his most important contribution was giving me reassurance on one point. I had written a chapter on "Chimpanzee Daughters and Jewish Mothers," and I had anxiety lest this offend my mother, of whom I am fond. Mr. Jacobs reassured that ". . . mothers are always treated badly in books, and I am sure that your Jewish Mother would rather be your Jewish Mother than Phillip Roth's." I hope I have not dealt too harshly with Jewish Mothers, for it is a personality pattern that may occur whenever women are oppressed, whether by religion, state, family life, or any form of minority status. I am grateful, for Jewish Mothers have consistently given the world its greatest psychologists. Freud, Adler, Perls, Bettelheim, Rank, Maslow, and Goldstein are but a few examples.

I owe a special debt to friends in the Human Potential Movement and the American Academy of Psychotherapists. Some of you were available when I needed you as teachers or therapists; I treasure from others moments of warmth and friendship when we enjoyed Lucy stories together. It was your enjoyment of Lucy stories that helped me decide to include my feelings and family life and not to write about Lucy with the rigorous emotional sterility with which scientific findings often are reported. In addition to George Brown and Judith Brown, I am grateful to Seymour Carter, Janet

Lederman, Koleman, Niel Lampier, Debbie Carson, Joan Fiore, Betty Fuller, Sheldon Kopp, Charlotte Selvers, Charles Brooks, and Natasha Mann.

I had superb editorial help from Peggy Granger. She gave me a warm, humorous friendship which dissolved the threat of editorial criticism, and the questions she raised were invariably thought-provoking and helpful.

Lucy at about age three being tested on ability to discriminate form.

INTRODUCTION

I am a psychotherapist and my daughter, Lucy, is a chimpanzee.

Nonetheless, while growing up, Lucy never saw a chimpanzee herself except on television or in a magazine. Although deprived of contact with her own species there were compensations, for Lucy had a chance to meet and befriend such distinguished students of chimpanzee behavior as Jane Goodall, Emily Hahn, Allen and Beatrice Gardner, Roger Fouts, Adrian Kortlandt, William C. McGrew III, and Caroline Tutin. And Lucy has seen pictures of herself in *Life, Psychology Today, Los Angeles Times, Parade, Science Digest* and *The New York Times*.

The reason Lucy never saw a live chimpanzee between infancy and adulthood and the reason she acquired such a distinguished circle of friends and acquaintances is that she was raised in a human home as a human child.

For almost ten years Lucy and I lived on a small acreage in the southwestern United States with my wife, Jane, and son, Steve. We lived about five miles from a small town and thirty miles from a large city, so we were isolated enough to have minimal contact with people who might object to a chimpanzee neighbor.

Jane took Lucy away from her mother on the morning of the second day of her life. Since that day almost ten years ago Lucy has been a member of the family. She ate with us, she slept with us, she worked with us and she played with us. In every way we could manage we gave her the same enriched environment that we provided our son, Steve.

This book is not a scientific study of chimpanzees; rather, it is a clinical or case study of one chimpanzee, raised under unique conditions and written from the point of view of one man—a psychotherapist. Both the observer and the observed are unique and generalizations to either species should be made only with extreme caution if at all. Yet I think the book will be exciting to the scientist as well as the general public, for what one chimpanzee did, others are capable of doing, under comparable conditions. Furthermore, what one man experienced may to some extent reflect the experience of mankind, if it were possible for mankind to live so intimately with our closest biological relative.

I have tried to state my biases, or more properly my feelings, as explicitly as possible to separate them from my observations and from inferences about what was going on in Lucy's mind. I hope you can see what Lucy did, when and where she did it, how I felt about it, and what I think her action means to both of us. Indeed, it would have been impossible to do otherwise, for Lucy formed an important part of my emotional life. She was as much in my heart and mind as was my son, Steve, or my wife, Jane. She influenced me enormously, far more than I influenced her, because her behavior is probably less modifiable through experience than is mine, since as a human being I am higher on the evolutionary scale and in a profession which emphasizes change and growth. Therefore, to tell my daughter's story I had to write about myself, and I made a commitment to myself to write with complete candor, regardless of how embarrassing to me it might be. This I have done.

LUCY: GROWING UP HUMAN

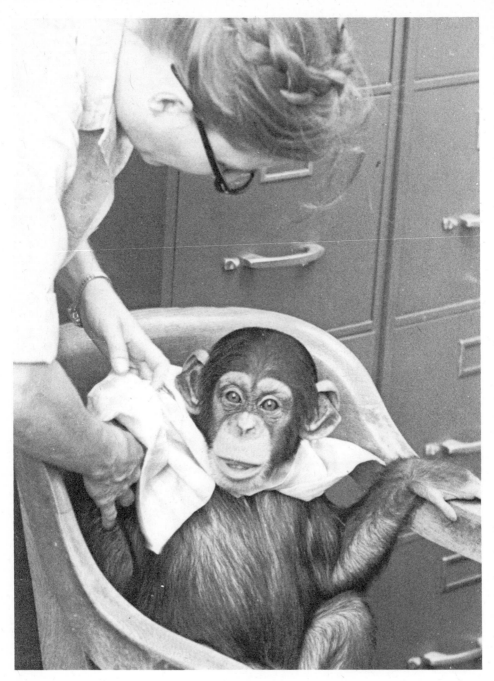

Lucy at the age of three being dressed by her mother.

CHAPTER ONE

Lucy Comes Home

"I AM A man; therefore nothing human is foreign to me."

"I am a man; therefore nothing animal is foreign to me."

When Jane and I decided to adopt Lucy and raise her as our own daughter we wondered what she would be like, *Growing Up Human*. Would she learn to love us, and perhaps have other human emotions as well? Would she develop personality characteristics by identification with us, as do human children? Would she be well-behaved, rebellious, intelligent, or stupid? And how intelligent might she be, given an enriched environment filled with opportunities to actualize whatever potential she brought from her chimpanzee mother to our home? And what about sex? When she matured sexually would she lust after an unknown chimpanzee lover, possessed of longing for an organism she had never seen? More likely we thought, but did not know, her sexuality would be shaped and molded by her experience, and be directed toward humans. And if then impregnated by artificial insemination, would she mother her offspring or abandon it, never having been mothered by a chimpanzee herself? Could she learn to talk? Could she learn to understand our speech? Would she develop a concept of herself?

We had questions about ourselves, too. What kind of chimpanzee surrogate parents would we be? I had a Freudian sexist bias at that time and I felt Jane would be a fine mother because she had endocrinologically-determined "instincts" for motherhood. But I had many doubts about my ability to become a good chimpanzee father. Indeed, I was having enough trouble being a good father to Steve. In Lucy's early years with us I often irritated Jane by asking for reassurance that I was being a good chimpanzee father. Of course, that was before I had freed myself of sex-role stereotyping with the help of Lucy and several therapists, (I almost wrote "other" therapists) and learned to do what came naturally without worrying about it. But at that time, "Mother knew best" and Jane was the ultimate authority on Lucy—which gave her much more responsibility than was comfortable.

I was aware of the scientific questions which might be answered by our raising Lucy, and also that her *Growing Up Human* might raise as many questions as it answered. Thus from a scientific point of view, it was a fine undertaking. I wish those had been my only motives; they were not. I also got involved because I was acting on impulse without considering the long range consequences to Lucy, to me, and to my family life.

Even our experience with Charlie Brown didn't deter us, although perhaps it should have. I think I should share this with you before Lucy "comes home."

About a year before we adopted Lucy we had acquired a three-year-old male chimpanzee named Charlie Brown. I had at first thought of him as an exotic pet, but he quickly endeared himself, so much so that he soon was a member of the family. He had a status and an importance to me I had never anticipated.

Charlie had an amazing ability for imitating Jane, Steve, or me—putting on our clothes and strutting about the house in a parody of being human. Sometimes he could even improve on our behavior, at least in the environment to which evolution had adapted him.

For example, we would go into the woods on our acreage and climb trees with Charlie. This was a source of delight to me. When I was a boy I had been stimulated by the novels of Edgar Rice Burroughs about *Tarzan of the Apes.* Tarzan fantasies had been my most delightful fantasies when I was a child, and they constituted a frame of reference which made it easy to believe that adopting Lucy was a good idea. And Steve, who was eight or nine at the time, and loved to climb anyway, would be supremely happy when Charlie was along. Charlie Brown would amaze us by his foresight. He would break off dead limbs before he would jump freely from one limb to another. The minute the three of us would start to climb a tree, Charlie would run throughout the tree, even on the highest limbs, breaking off the dead branches and throwing them to the ground so that we could play together with no danger of being hurt by a dead limb breaking beneath us. In three or four weeks time Charlie was the fascinating center of our life. He did everything with us. He was so well behaved that frequently we would dress him in children's clothes and take him to the office or to town shopping, where everybody loved him.

Charlie Brown died at the age of four as a result of a freak accident. He looped one end of his security blanket through the cross-pieces of steel at the top of his cage, where Jane and I had left him for a short time. He could then swing back and forth within his cage, but somehow he got twisted up in the blanket while swinging and died of suffocation. We consoled ourselves by rationalizing that Charlie Brown had had a happy life, if a short one; and that he had died quickly while playing; as good a way to die as any. But the fact is that I was griefstricken and it was years later before I completed what Freud called "the grief work." In retrospect I am sure that one of my unconscious motives for adopting Lucy was to help me come to terms with the loss of Charlie Brown.

So it was thoughtless of me, although human, to go along with Jane and adopt Lucy only three weeks after

Charlie Brown died. Instead of avoiding further involvement with a baby chimpanzee, Jane and I went in the other direction. We committed ourselves totally to Lucy. We determined to take no chances, to spare no expense, to devote ourselves completely to raising her as our own child.

Jane Goodall found when observing chimps in nature that each chimpanzee had a distinctive and unique personality, and our experience with humanized chimpanzees was much the same. Charlie Brown and Lucy were quite different, though they had one wonderful characteristic in common: they took themselves and the people they loved for granted, totally trusting once a relationship of love was established. When two human beings love one another the relationship is fraught with risks and only the most well-integrated and courageous people are capable of loving. Several examples may make clear what I mean. Everytime I have loved another human being there always has been a moment of risk during which I felt I might not be loved in return. Or, since I loved a a particular person, that person became extraordinarily important to me and acquired power either to cause me pain or to bestow great pleasure and satisfaction. When I was young I sometimes feared to love because the loved one might have power over me, controlling and manipulating me as a puppet on strings of guilt and obligation—"If you really loved me you would . . ." The rewards of love have always made it worthwhile on balance, but it always has had risks and pains for me.

But with Charlie Brown and Lucy there were no risks to loving on either side. Once we had established a relationship of love and trust, their responses to me and mine to them were consistently warm and loving. Once or twice I felt that Jane loved Lucy or Charlie Brown more than she loved me, experiencing a kind of jealousy often reported by men in psychotherapy who feel rejected when their wife devotes most of her time and energy to the new baby. But I never once felt that either Lucy or Charlie loved Jane or Steve more than me, so consistently warm, loving and grateful were their responses to me. They would get mad at me, of course; life is too complex to be without frustration, even among the loving. They would scream in rage, every hair erect, lips pulled back, teeth exposed, pulling their hair, pounding the walls—for all of 30 seconds! Every child has had similar experiences if he has grown up with a dog or a horse, for although dogs and horses are very different from chimpanzees, they may be loved without the reservation or ambivalence of loving another human. Ask yourself when you suffered more pain? When your dog was run over by a car or when a parent died? The answer for many people can be embarrassing but instructive.

Back to Lucy

We had been thinking of adopting another chimpanzee, so we became very excited one day when Jane heard of a captive adult female who seemed to be pregnant. This was especially interesting because to adopt a newborn chimpanzee would mean it had no time to learn chimpanzee behavior, and could be more completely humanized. The lead proved to be a good one. Jane made all the arrangements and flew to the East Coast on the day the infant was born. In the early morning of her second day Jane fed the mother a Coca-Cola which had been spiked with phencyclidine, a drug which puts chimpanzees into a deep, pleasant sleep. When the mother fell asleep the handler entered the cage, took the baby from the body of the mother, wrapped her in a baby blanket, gave her to Jane, and our adventure had begun.

Jane named her Lucy and brought her home on a commercial airline, carried in a bassinet, her face covered with a lacy blanket. There were several humorous incidents on the plane as well-wishers wanted to see the baby, and a few tragic moments as Jane took the baby from her mother. But those are Jane's stories and await her book.

We were blissfully unaware of the complexities we were creating on the day Lucy came home.

We were living in a small, two-bedroom stone house situated at the northern end of five acres. It was a pretty good place to raise a chimpanzee since at that time the city was five miles away, and urban sprawl had not caught up with us. Thus we did not worry about neighbor kids coming around with BB guns and rocks to tease Lucy—eventualities we later had to handle. The outside of the house was of stone and that part was durable, but that was only the outside. The inside walls were of painted sheet rock—no barrier to a chimpanzee hand, and the windows were casement windows which opened by turning a handle, rather than being lifted up or down; no problem at all for Lucy.

There was a bedroom for Steve and one for Jane, Lucy, and me. We planned that Lucy would sleep in a regular baby crib in our bedroom with us, because at first she would need regular night feedings, and Jane could be close at hand at all times.

The house was perched on a hill, and on the east there were some big trees which we knew that Lucy would someday climb. When she was an infant, however, these were too big as there were no limbs near the ground. In the front of the house there was a concrete fish pond by a large clump of bamboo and at the west end of the pond was a medium-sized mimosa tree which was to be the first tree that Lucy was to climb. She was scared to death when, at the age of 14 months, Steve held her on the lowest branch. He had to reassure her for some time before she would even sit 12 inches off the ground without screaming in terror.

Sixty feet from the back of the house was a small concrete building twenty-four feet long by sixteen feet wide. I used it as a combination bird laboratory and shop. All my adult professional life I have been interested in breeding rare parrots; their beauty and intelligence fasci-

It's a big world out there when you're a baby chimpanzee in a human home.

nated me, and trying to breed them appealed to the conservationist in me, as many species are becoming very rare in nature.

I always had at least a dozen parrots, macaws or cockatoos around the house; sometimes more. Incidentally, we were determined to keep Lucy out of the bird lab because that is where Charlie Brown had been when he accidentally hanged himself.

When mother and daughter reached home, Jane was exhausted, apprehensive, concerned about Lucy's mother's loss, and filled with doubt about her ability to meet the needs of this extremely small and dependent infant.

In my opinion, the airplane flight and the act of taking Lucy away from her mother had been for Jane the symbolic equivalent of the act of giving birth, and formed between them as close a union as the bond between any human baby and its mother. Indeed, the symbiotic union between Jane and Lucy was much closer, for the infant chimpanzee requires twenty-four-hour a day contact with the mother's body for the first year of its life. During that first year Jane tried to approximate chimpanzee mothering as closely as possible.

Baby chimpanzees are far more intriguing to me than human infants at birth. Human infants at birth all look pretty much alike. Probably few human mothers will agree with this statement. But if one's own child is not being compared the situation is different. Imagine looking at a line of ten newborn infants, all wearing identical diapers, in the same kind of basket, in a maternity ward of a hospital. They all have pretty much the same faces, all moving with mass uncoordinated motions in the same amount and in the same way. This is not so for chimpanzee infants. From birth they are far more active motoricaly than human infants. They begin to explore their environment more quickly and they have black wrinkled faces which are never expressionless. Skills such as crawling, walking, and climbing develop much faster than in the human infant. Thus the proud parental observer does not have to wait long before he can see his beautiful daughter do something clever and wonderful which fills him with delight.

Infant chimpanzees are very appealing. They are smaller than human infants at birth. As I looked at Lucy I never realized that in eight to ten years she would weigh eighty-five to ninety pounds and be five to seven times as strong as I was. Both men and women would look at Lucy's responsive brown eyes and wrinkled face and be captivated and charmed from the outset. She was so responsive to being looked at, held, and stroked. Even men who did not think of themselves as being maternal, or as even liking children, sometimes found it hard to resist picking Lucy up and holding and cuddling her. And the infant Lucy would bring out maternal urges in young unmarried women, even counter-culture types, quicker than anything I could imagine. There was, however, exceptions; some people could not imagine what we saw in such a funny-looking creature. They were a minority.

One experimentalist had a different reaction, one I will always remember. He looked at Lucy and smiled, mixed himself another generous drink, and casually suggested the most gruesomely horrifying experiment. I remember it in the same way I could never forget "the banality of evil," to borrow Hannah Arendt's phrase. He asked what we were doing with Lucy, and when I said we were raising her as our own daughter, and giving her all the enriched contact with human beings and culture that we gave our own son, Steve, he interrupted with, "Wouldn't the results be contaminated by her relationship with you when she grows up? Wouldn't it be better to raise her in total isolation and feed her by machine? Then when you study her behavior in adulthood it would not be contaminated by the relationship she has had with you?" Actually, chimps have been raised in total isolation before. Even two years of isolation is enough to produce aberrant behavior.

Lucy's writing is much less messy than her fingerpainting but it's not nearly so much fun.

One chimp so raised was not a chimpanzee except in the biological sense. He was terrified of chimps, people, and many inanimate objects. The only thing to which he seemed attached, and which turned him on sexually, was a 55-gallon oil drum.

But anyway, to most people, Lucy was a little bundle of cuddly love with a wrinkled black face and four exploring hands which would reach out in all directions at once while she lay in her crib on her back. She could grasp with either hand or foot with great dexterity. She quickly learned to hold her own bottle. Jane would fix her formula, which roughly was the same as the formula fixed for human infants at her age, and she had no feeding problems whatsoever.

She smiled (what we called a smile; it really is a grin) and laughed from the first day; this seemed to be her undifferentiated response to any kind of stimulation, social or physical, when she was not sleeping. She was, of course, sleeping most of the day, but usually stayed awake smiling for about a half hour after each feeding.

At two months her eyes would focus and follow a bright object, and at three months she could walk about on all fours. At five months she was trying to climb out of her crib to go to people, and at six months she was pretty mobile on all four limbs, though she did not walk on two legs till she was about eight months old.

She had a concept of play and a "play face" at six months, and would laugh during tickling and would love to tickle and be tickled; characteristics she still has. She would walk alone, but never very far from Jane or me at eight months. We would be sitting on the lawn, for example, and Lucy would at first go two feet away and that would be a critical distance, and she would play only within that radius of her mother. Then that critical distance gradually expanded to three, four, then five feet. There were times when she got so involved playing or exploring that she would exceed her critical distance, then realize how far away she was, scream in terror and rush

*This sequence of Jane and Lucy and her coloring book
show the tender and loving bond between Jane and her
unusual daughter.*

back to Jane. She was always very affectionate; perhaps because her needs were met and, having never been punished in her early childhood, she knew only warm loving human contact. We tried to follow the example of chimpanzees in nature, to teach Lucy by loving example, not by punishment. The only punishable offenses were biting someone or running away, and these never occurred until much later. Then punishment was by threat or verbal abuse—much sound and fury, but never physical punishment. We did not think physical punishment necessary, or that it would be productive of anything but more hostility. So from three or four months on she was kissing and being kissed by everyone who wanted to exchange a touching of lips and interact with her in this way.

By the time she was a year and a half old she was climbing all over beds, tables, chairs, and cabinets and it was clear that we would have to modify the house to control her for her own protection. We then put locks on our kitchen cabinets to keep her away from household poisons, locks on our bathroom doors and cabinets for the same reason. Eventually all our internal doors had locks.

While young she had several respiratory infections, presumably those that human children have, although of course, we could not diagnose them as precisely. But she would have a runny nose and fever. When she was sick we would usually know it by her reluctance to play at first, and then much later by a loss of appetite. We became sick with anxiety at these times really not knowing what to do. We treated her entirely as a human child and insulted many pediatricians whom we called and asked for advice. They usually wanted to refer us to veterainarians who, of course, had been trained to understand the physical problems of cows, sheep, and goats but not of chimpanzees or humans. Fortunately, we were able to find several competent pediatricians who were not worried about practicing veterinary medicine without a license, and who were of enormous help when we needed antibiotics or some equally critical medication.

Table Manners and Hand-Me-Downs

Before long she was eating at the table with us. We did not have to teach her to use spoons or forks. She would see us using silverware and immediately do so herself. There seemed to be no trial and error whatsoever in learning to use table silverware. By the time she was 15 months old she seemed quite comfortable with all the paraphenalia with which humans consume their food. Since that time, she has always used glasses and cups for drinking, spoons for eating. Similarly with silverware. If we just hand her a bowl of food, she will go to the cabinet and get a spoon or fork and use it appropriately, and has continued to eat this way.

Lucy used a highchair she inherited from Charlie Brown, even though a highchair had not worked very well with him. For example, when Charlie Brown used the highchair there were many accidents because he was so agile and strong, and he had been more aggressive than Lucy. The highchair really would not hold him and support him during his meals as well as it would a human child the same size; he was simply too strong and too active.

One incident with Charlie Brown in the highchair stands out in my mind. Indeed, I shall never forget it. It was one evening when Charlie seemed to be unusually hungry, and perhaps a bit irascible. He asked for his dinner thirty minutes early by making good-food sounds, panting, "uh uh uh," and climbing into his highchair. Jane got the message and began to prepare his dinner, but apparently she was not fast enough, for Charlie began banging on the metal tray of the highchair, and crying a rather shrill scream. She was fixing oatmeal to which protein supplement and vitamins had been added, and a Charlie watched Jane working on a kitchen counter four feet away he began to become more and more agitated. He threw his spoon and cup on the floor, and when I picked them up and put them back on the highchair counter, he immediately knocked them off with the back of his hands. I was preparing a barbecue sauce as I planned to cook some ribs after we got Charlie to bed. The basic ingredient of the sauce was catsup to which I was adding worcestershire, tobasco, garlic, lemon juice, and honey—all highly colored and sticky ingredients which I was mixing in a bowl. Standing on the counter beside me watching the whole proceedings was Cocky, a tame Moluccan cockatoo, who thought she was human, having been imprinted on humans from an early age. Moluccan cockatoos are among the most beautiful of the cockatoos. Cocky was twenty-two inches long with a pinkish white body, a ring of blue around her eyes and a salmon pink crest which she erected whenever she was excited. The crest was erect now as Cocky stood beside the mixing bowl on the counter watching Charlie become more agitated. Charlie then went into a complete tantrum. He screamed at the top of his lungs, beat his hands against the table, and tore out great chunks of his hair. Jane, determined not to be manipulated by his tantrums, told him, "Go ahead, tear your ears, that's good, too!" Charlie screamed louder and, unable to contain himself any longer, jumped from his highchair four feet through the air toward the bowl in which Jane was mixing his dinner. But he was so excited that he miscalculated and landed in the middle of the barbeque sauce, which splashed all over my gorgeous white cockatoo, soiling her a bloody catsup red. I became sick on the spot—sick of the mess; sick of Charlie and speechless when I looked at the hideous mess that used to be my beautiful pet cockatoo. By now Charlie, of course, had his face buried in the bowl of cereal, ignoring everybody. Jane, ordinarily the tenderest of women, suddenly became a monster. She grabbed Steve's lever action BB gun which was on top of the refrigerator, pointed it at Charlie, and in one motion, cocked it and said, "You long-haired son-of-a-bitch, get back in that chair or I'll blow your guts out!" Charlie ran toward the living room and Jane fired one shot from the hip to scare him. It missed him and broke a window, but that was enough to scare Charlie back into his good behavior. He ran back, climbed into the highchair, assumed a virtuous facial expression and began to whimper. I went outside with Cocky and a damp cloth and tried to wipe off the sauce; she was, however, never completely white again until after her molt the following year. The BB gun became a symbol for Charlie though, and after that incident we never had to use it again. We simply carried it with us whenever we went outside with Charlie, and from then on he was very well behaved. Before that incident, however, if we were playing outside and it was time to go some-place, Charlie Brown would climb the tallest tree and laugh at us. This would be very frustrating if I had an appointment or a class to teach, because I just could not go off and leave him. I would call, plead, and threaten—in that order—and he would just ignore me, or act as if he had not heard me, or when I would curse and threaten he would break off a few small limbs and toss them down, and I would feel impotent and helpless. But the BB gun incident changed all that. Charlie would see me carrying the gun, and for him fear was the mother of virtue.

The first three years with Lucy were much easier than with Charlie. Lucy was much quieter, much less aggres-sive. She was gentle and curious and cuddly where Char-lie had been boisterously and aggressively playful. She was such a "good" infant we could take her everywhere.

Jane would take her to her office each day. She at that time worked as an administrative secretary in a psycholog-ical clinic, and Lucy would sit in a chair, covered by blankets, or play in a corner on the floor, with blocks, dolls, small toys of all sorts. She was quite curious and just about anything could become a toy for her, regardless of its original function. She could take the nipple off her bottle and use it as a pacifier or teething ring, and the bottle would become a delight to roll across the floor.

She would get a bottle in the morning and again in the afternoon wherever she was, and then would take a nap. If she was hungry or thirsty she had no trouble indicating it. She seemed not to whimper or cry as much as a human child the equivalent age, but when she did she could quickly be comforted by picking her up. Everyone liked to pick her up. Often men who I thought were male chauvinist stereotypes and would never think of holding a human baby to their shoulder would ask permission to pick Lucy up and comfort and hold her.

She was so adorable everyone had about the same reaction to her. Steve, for example, tried to avoid getting involved. He said (and he was right) it was too soon after Charlie to get him a sister, because he still missed Charlie. Yet he was quite attached to Lucy within one day. He would cuddle her, carry her on his shoulder, and care for her in a maternal way at first, then as playmates when she got more active. She was so intrinsically interesting Steve apparently never felt sibling rivalry. Steve loved to show her off to his friends. He was the only kid around with a baby chimp sister, and his friends would visit and admire her. And she would ride in the car with us when we went to town, to take Steve somewhere, or to shop. When she was two we could even take her into grocery stores, and she would sit in the cart as we shopped up and down the aisles. Occasionally a manager would say, "no pets allowed" and we would become indignant and say that Lucy was not a pet, that she was our daughter. In most cases that sufficed and the skeptical manager would shrug helplessly, though once or twice we did have to leave a grocery store.

Lucy never misbehaved. We never had to put her on a leash until she was six or seven; she would be very comfortable and well-behaved in a car, office or store as long as she was close to one of us. Of course, she was much closer to Jane than to Steve or me for the first three years.

At home she was much more active. She felt more secure in the house, where she had her own crib, playpen, toys and where her whole family gathered each evening. Generally Jane fixed Lucy's dinner, a mixture of Gerber's baby foods (fruits, vegetables and meats) warmed just right, tested for temperature against the lip, though sometimes Steve and I would feed her. She had her own silver baby spoon and cup, which Steve had used, and sitting in the highchair she would often sweep it to the floor with a loud satisfying crash, as Steve had done before her.

She gradually became more "aggressively" active, learning quickly to climb doors, flip light switches, open cabinets, turn water on (rarely off), take books off shelves, and, in general, to act like a small, lovable cyclone.

Chaos

By the time Lucy was three years old we knew we were in trouble. The family was chaotic to the point where an explosion was imminent. Lucy was into everything; she could take a normal living room and turn it into pure chaos in less than five minutes. She would scurry across the mantle above the fireplace like a black streak, knocking knicknacks to the floor; unscrew the light bulbs, dropping them to the floor and breaking them; climb a lamp and jump from the lamp to a fixture on or near the ceiling; take books out of the bookcases, look at them for a moment and discard them; take pictures off the wall, study them intently on the floor; roll glasses back and forth on the floor; run to the refrigerator, take out oranges or apples or any kind of round fruit and play ball with them, running rapidly across the floor rolling the fruit; or she would disconnect the toilet paper, hold one end and run laughing through the house, leaving streamers of paper behind her. Her playpen would not hold her and, partially because Charlie had died in a cage and because Lucy seemed so human, we were determined not to put her in a cage. I had not yet learned to deal effectively with my own anger, and could not express it toward Steve or Lucy, even though it would have set limits for them. Later when I did learn to explode without guilt, I had no more problem controlling Lucy. I think Jane felt Lucy was "so cute"—and besides, as Jane put it, she was the one doing the cleaning up after her, not me, so why should I object? But I did object, yet if I yelled at Lucy, "Goddammit, sit in that chair and be still!" I felt guilty afterward. (My parents had never been able to control me when I was a child, either.)

Lucy would go into the cabinet, unscrew a jar of honey, take a few swallows making good-food sounds and then lay the honey jar down sideways on the carpet. I would be horrified when I saw the mess while Jane would say, "Did you notice how she unscrewed the lid all by herself?"

Frankly, I was beginning to feel discriminated against. Perhaps I am overstating this, but it's the way I felt at that time. Lucy could tear the kitchen apart and Jane would say, "Isn't she clever? Isn't she wonderful? Notice how active she is!" But it seemed to me that if I were to make a cup of tea and leave the tea bag on the counter, she was likely to say, "Look, I'm not your mother, it's not my job to pick up after you!"

Steve and Lucy were getting along great and Steve was not a part of the mess hassle. They would go outside together, play in the trees; Steve and I built a treehouse and he and Lucy would play in it for hours and give us some privacy and peace. But in the house the control problem was becoming intolerable. We needed a room for Lucy. She had to have a large room of her own with toys and a trapeze for exercise or something comparable so that we could lock her up from time to time in comfortable surroundings to which she would not object. This led to further conflict between Jane and me because I felt we could not afford either a new house or to build an addition onto our house. Jane felt we had to do it for Lucy's sake. Certainly the family would have exploded without more room, and there was no way to complete our study of Lucy and raise her to maturity without it.

We were deeply in debt at the time; I knew we could not afford to build onto the house, yet I knew we had to do so. By then I loved Lucy as though she were my own daughter, and knew we had to do something for her. My relations with Jane were becoming more and more strained. Also I was not liking my job as chairman of the psychology department, and felt I was losing control over my life.

Somehow, out of all this chaos there came a plan to build.

Lucy's House

Since it was a house for Lucy it had to be built out of the strongest possible materials—stone, brick, concrete, or steel. Since concrete was the cheapest commensurate with human comfort, we decided to build the walls out of reinforced concrete block and the roof out of prestressed concrete beams. The floor, too, was to be made out of concrete. We wanted Lucy's room to be away from the living room and the bedroom, as a chimpanzee can make an enormous racket, yet close enough so that we could see and hear her. A full-grown chimpanzee can lie on its back and kick with its feet against the wall until an unreinforced concrete block will crumble. So Lucy's room was of steel reinforced concrete, large bedroom size, fifteen feet wide by eighteen feet long added onto the northeast corner of what now was the "old part" of the house. In the roof of Lucy's room there is a steel trap door which she can open at will which allows her to enter a large L-shaped room on the roof. So Lucy's part of the house really has two stories—a large room on the concrete roof and a large bedroom below. The doors, of course, were all steel doors and the one to the outside was a double door so that equipment could be brought into her room for cleaning.

Maury and Lucy enjoy fishing in their pond. Note sturdy construction of Lucy's rooftop apartment.

In the center of the house we built a large courtyard with a fiberglass roof. I had been to South America a number of times for the Peace Corps and had seen many houses built about a central courtyard. Since we were building onto a rock house we really only had to build three sides before we had a four-sided enclosed structure. So now in the center of our house there is a courtyard twenty-five feet square with a concrete floor in which there are floor drains, as there are in Lucy's room. In our courtyard there are all kinds of trees; since the ceiling is sixteen feet high, some of the trees are pretty large, and a collection of orchids, Bromeliaceae and palms. I once had a large banana tree but Lucy completely destroyed it. We have several orange trees and a lemon tree growing in the courtyard, but it is almost impossible to grow fruit to maturity as Lucy grabs them and eats them even when they are still very small and green. The roof of the court-yard is made of two layers of fiberglass separated by an inch of air space for insulation so that it is really not too difficult to keep it warm in winter. The living room faces to the south. It is large—thirty-nine feet long by twenty-five feet wide—enough for Lucy to run around and around the furniture, playing chase and large enough to hold a reasonably good-sized seminar or therapy group in. At the northern-most edge of the living room is a wood-burning fireplace and a large picture window through which can be seen a triangular-shaped terrace leading to a pond, approximately two acres which laps at the base of the terrace. All of us, including Lucy, have enjoyed the pond, which was the result of a comedy of errors. We had asked an architect to design us an addition which would be adequate to house Lucy, be reasonably comfortable for humans, and within our budget. When we finally approved the plans and put them out for competitive bids, the lowest bid was approximately $15,000 above our limit, so we had either to abandon the plans or to cut everywhere. So, we cut everywhere and one of the "cuts" created the pond. The addition was going to be built on a hill, and

$2,000 worth of river sand was necessary for leveling the site before construction started. So we saved approximately $1,500 by building a pond for $500 and leveling the site with the dirt removed to build the pond. Then we filled the pond with water from our well. It has been a source of great pleasure as well as food because we stocked it with channel catfish. Since catfish are omnivorous we put into our kitchen an extra sink with a garbage disposal unit and piped that into the pond. In some years we have grown as much as 1,000 pounds of edible catfish from our own table scraps. The front porch of the house meets the water's edge and all of us have enjoyed sitting there and fishing. In some years the fishing was so good that we could invite people for a fish dinner with no fish in the freezer, secure in the knowledge that we could catch them off the front porth thirty minutes before dinner and, quickly cleaned, they make a delicious dinner because they were fresh and fed on a human diet. Lucy loves to "fish" and she has learned to reel in, but she cannot cast very well and usually gets a backlash. She screws up the reel to the extent that I usually do the casting for her, though she can reel the bait in and has several times caught her own fish. She seems fascinated by its slimy wiggling body as it flops helplessly on the porch, poking at it with her forefinger for a moment, then quickly losing interest. It has always surprised me that she is so agile. She has never hurt herself from the fins of the catfish. Though I have fished since early childhood I frequently get stung by a fin while taking the catfish off the hook or cleaning it.

On the east side of the living room are four large windows through which may be seen six bird flights. In each of these flights there is a pair of cockatoos, usually Moluccan cockatoos, as these are my favorite birds. During the warm months they can be seen flying back and forth in their flights, Lucy loves to "devil" them. When they are perched near the living room window she will sometimes charge the window and bang on it, startling them until they fly to the perch at the other end. They can enter a window at the far end of the flight into an indoor cage which is climate-controlled, and during the winter usually are inside.

The furniture is held to a sparse but durable minimum. When Lucy went through a teething stage we discovered that human furniture is not chimpanzee-proof. She bit through leather, naugahyde, or any other fabric that we could find. Finally we settled for two "Danish modern" sofas and two Barcelona chairs with one large coffee table in the middle. They aged rapidly under Lucy's use, as she builds nests of the sofa pillows, arranging them about her in a circle on the floor, then playing in her nest.

In the kitchen we built an island for the range vented through a large brass hood in the ceiling. It made it very convenient for Jane and me both to be in the kitchen at the same time without getting in one another's way, but it was also a great advantage for Lucy because she can run around and around the island at great speed, or hide from us. She can move so softly through the kitchen that we cannot hear her.

Just south of Lucy's room and connected to the living room by a corridor is my office. I used to see individual psychotherapy clients there and still do sometimes, but mostly now it is used for groups.

Since we are in the middle of five acres, we can look out any window and see trees, some of which we planted, while others are natural. So we get the feeling of being isolated even though we are only five miles from downtown, and there are neighbors four hundred yards in any direction. This really has bothered us every time Lucy runs away. Once, for example, she got out, ran about half a mile, entered a neighbor's house and raided the refrigerator. They were a retired couple, and while they were frightened, they were very nice about it, because it added some interest to their life in their declining years. Now whenever their grandchildren visit they bring them by to see Lucy. I am grateful for this because they could have shot Lucy as a burglar and sued us.

It took about a year of planning before the architect, Jane, and I agreed on a set of plans. The builder said he could do it in three months. One year later we settled with him with many details unfinished. When it was finally finished it was great. Everybody who saw it thought it was beautiful, and so did we until winter came. Then we found that concrete blocks, even with the holes filled, were very poor insulation and the house was always cold. Furthermore there were such wide expanses of concrete that the acoustics were terrible. The living room had a twelve-foot ceiling—the architect thought that concrete beams twelve feet from the floor in such a large living room would give us a feeling of "great space and openness." Maybe so, but it cost a fortune to heat. It was the floor that caused Jane and I our greatest difficulty. Lucy was not toilet-trained.

Toilet Training

I have heard half a dozen experts say that chimpanzees can be toilet-trained, but I do not believe it. I have even heard Jane say that one particular infant chimp that she saw placed in a human home was toilet-trained, but with all due respect, I do not believe it. I saw that particular chimp, and though he might have been toilet-tained, he impressed me as being over-trained to the point of being an obsessional neurotic with impulse and constipation problems. Be that as it may, we have never been able to toilet-train Lucy very effectively. It is not that she cannot understand the tool function of the toilet. Indeed, she went through a period of being fascinated with the toilet as human children often do on a magical "now you see it now you don't" basis. But she will not use it consistently. If we tell her, "Lucy, go to the bathroom and use the toilet," and if she needs to go to the toilet, and is in a particularly compliant mood at the time, she will go to the bathroom, sit on the toilet, use it properly, and even flush it. But she is always careful to retain part of her feces and urine so they may be deposited on the living room floor a moment later. She sees us use the toilet all the time and she frequently will voluntarily hop up on the stool and use it herself—to evacuate fifty percent (approximately) of the then available urine or feces, so the balance may then be deposited in the most inappropriate inaccessible, and difficult-to-clean place.

In our living room we once had a large glass dinner table. The top of the table and the legs were both made of plate glass. The table was designed by the architect who designed Lucy's addition to our old house, and it was a quite lovely and striking table. Since it was glass and the floor was concrete it had to sit on a rug. So at one end of the living room we had a large green shag rug with the glass dining table in the center of it. Three-fourths of the living room had a painted concrete floor—as good a place as any for Lucy's toileting, if she was not going to use the toilet. Since it is easily cleaned it would even be the preferred spot, I should think. No so, however; Lucy invariably chose to urinate and defecate on the shag rug beneath the glass table.

To put it mildly, I have never liked this behavior. I have an obsessional streak in me, and I prefer an organized world in which feces are deposited in their proper place. I have not, however, been able to communicate this to Lucy. I have taken her to the toilet countless times, whenever she needed to go, and when she did not; she complies with the letter but not the spirit of my injunction, always reserving part of her bowel or bladder contents to deposit them elsewhere, as said before—no matter how long I keep her sitting on the toilet. I have cursed, threatened, pleaded, and tried to bribe her to no avail. The only response to my pushing toilet training is that she sometimes tries to help Jane and me clean up her mess, which always makes it worse as she understands the form, but not the substance of scrubbing the floor.

As I wonder about our inability to toilet-train Lucy, I conclude that she refuses to control her bowel and bladder because it makes no sense to her to do so. That she could do it if she wished is demonstrated by her using it for half her movement, then retaining the balance to void it where she wishes. Also Lucy has never had an accident of any sort in the car, even on trips as long as two hours. Lucy simply has not acquired the values of the Judaic-Christian tradition in American life which hold that the by-products of our metabolic processes are dirty and taboo. (That they retain some interest for us in spite of social pressures to be nice and not nasty is evidenced by the persistence in our language of the middle-American colloquialism "fuck that shit.") To Lucy, urinating and defecating are natural and pleasurable acts and she is determined to do them where she wishes. If she cannot control her environment, at least she can control herself, and urinate and defecate in favorite places. While these conclusions have the scientific status of unverified hypotheses they fit the data and, like so many theories in psychology, they permit the prediction that under controlled conditions Lucy will urinate and defecate when and where she damn pleases.

I think Jane never really accepted, as the parents of retarded humans frequently cannot accept, that her baby was retarded in the area of toilet training. Lucy is a genius chimp intellectually, but she is "stupid" and rebellious fecally. Jane wanted to carpet the living room. The very idea makes me ill as I write it. I felt it was utter nonsense to carpet it with Lucy not toilet-trained. Jane felt the room was too cold and if we had wall-to-wall carpeting it would be an insulation, and that it would force us to work harder to toilet-train Lucy, who would then be toilet-trained. We argued and we fought, backwards and forwards. We couldn't afford it and it wouldn't work, but we needed it. In the end I gave in and we spent $2,000 to carpet the house. Six months later even Jane had to admit it was hopeless. There were smelly little brown stains all over the house that no amount of soap and water could remove. We discovered at this time that many of the products sold in grocery stores for reducing odors do not work, and many miracle cleaners were no better. Finally, the carpet came up and waterproof and fece-proof vinyl went down. That is the floor we have today. The vinyl was not as good an acoustical sponge as was the carpet, but better than raw concrete, and it did lead to an interesting experience.

Intruders?

I awakened about midnight to hear muffled noises in the kitchen. I lay in bed without stirring for a moment, trying to be sure. I again heard a soft clang, then silence; then a click and more silence, as though a drawer were being opened and closed or a plate lifted from a shelf. There were no lights on in the house. I quietly got out of bed, got my pistol and cocked it, and started walking slowly in a deep crouch toward the living room. The sounds continued and I was certain intruders were in the house. I sensed another presence, as I stood there, naked and scared, but I could see or hear nothing. I braced myself and extended one hand towards the light switch while I pointed the pistol in the general direction from which the sounds had come. Suddenly I felt a hand grab my ankle from behind. I spun around pointing the pistol at the floor just ahead, I hoped, of my ankle. It is a miracle that I did not fire before I realized it was Lucy grasping my leg. I flipped the light switch and the kitchen debris told the story. Lucy had awakened in the middle of the night, was hungry, and had gone into the kitchen to raid the refrigerator. This was my first introduction to Lucy's pattern of raiding the refrigerator in the middle of the night—a delight of civilization she appreciates to this day. She was not happy with her diet at that time, nor were we, and our attempts to wean her to a more satisfactory diet were to provide me with many tears, some laughter, and a dramatic lesson in self-awareness.

CHAPTER TWO

Chimpanzee Daughters and Jewish Mothers

IN NATURE the baby chimpanzee is enormously dependent upon its mother throughout a long childhood. Human beings and chimpanzees experience about the longest period of infantile dependency of any organism on this planet. This dependency is upon the mother exclusively as no chimpanzee infant knows his own father, because the female may copulate with any interested male whenever she is in estrus. The infant is so dependent it develops an anaclitic depression (or what would be a depression in humans) if the mother dies or is otherwise absent during this period. It may even die if the absence of the mother is too prolonged. This extreme dependency upon the mother, a characteristic of all chimpanzees, may be the reason Lucy was so susceptible to manipulation by Jewish Mothers. That is, her great need may have created a desire to please, regardless of time, place, or circumstance. In some situations, for example, Lucy wants to do her own thing and will not obey us unless the communication is couched in guilt-producing terms, for example, "Lucy, how could you do this to me?" But before I can go into this further, I want to say what I mean by "Jewish Mothers."

The Jewish Mother, as I am using the term, refers to a very primitive form of personality organization, the salient features of which are using guilt to manipulate, speaking in double-binds; using complex, sanctimonious rationalizations to avoid responsibility for one's words and deeds; and masking hostility with self-deprecation and obsequiousness. Sheldon B. Kopp, in his wonderful book on psychotherapy called *If You Meet the Buddha on the Road, Kill Him!* tells the following anecdote his mother told him to illustrate "a mother's love." A careful reading of it will reveal a mixture of sex, aggression, and hostility masked by sanctimony that is characteristic of Jewish Mothers when controlling through guilt.

Once, a long time ago, it could be anytime, a mother is a mother no matter when . . . So, once, at some time in the past, there lived a poor widowed Jewish mother and her son. By herself, she raised him. He was not a bad boy, but when he was eighteen, he fell in love with a shicksa. Plenty of nice Jewish girls there were in his village. His poor mother told him to stay with his own kind, but youth is deaf and blind and foolish. So . . . what can I tell you? These things happen. It came to pass that he fell in love with this beautiful shicksa, the unthinking love of the young.

His mother, of course, knew he was troubled (a mother always knows), but she did not know the reason because this foolish boy did not confide in her. Now the shicksa was not serious about the boy, fool that she was. She only played with him. What can you expect from a Christian head? At last she tired of him. This was his chance to escape back to his mother. But a boy, a fool, what he cannot have, that he must have. He told the girl he would do anything she asked, anything, if only she would marry him.

At last, to be done with him forever, the shicksa made a terrible covenant with him, knowing he would never fulfill it. Not even a goy would ask this in seriousness. But she did not know the fire of this boy's foolish infatuation. She told him . . . she told him . . . I can hardly bring myself to tell you . . . , she said to him, "I will marry you only if you cut out your mother's heart and bring it to me. Only this way can you prove your love." The boy was filled with horror. To kill his own mother . . . Yet, he must have this forbidden girl. And so he stole into his own house like a thief in the night. And in the dark of night he knelt beside the bed in which his mother slept the sleep of the good. He knelt and prayed that the Lord God would understand and forgive him for what he must do. And so, this ungrateful boy took from his belt a knife from his mother's kitchen, and plunged it into the breast of his poor sleeping mother. He killed her and cut out her heart. He could feel the warmth of her heart in his blood-stained hands as he rushed from their home to the house of his shicksa. As he ran up the cobblestoned street of their village, with the heart of his mother clutched in his guilty hands, he stumbled and almost fell.

And, as he stumbled, he heard the voice of his mother's heart speak to him from his hands.

His mother's heart said, "Be Careful, my son."

That is a mother's love.

The Jewish Mother is a very old and primitive form of personality organization, probably existing long before Christ. At that time Jewish women lived in a state of complete submission to the authoritarian male. "I thank Thee, O Lord, that I was not born a woman," is an orthodox Jewish prayer that reflects the attitude of master toward a somewhat defective slave in the emotional climate which creates Jewish Mothers. As Nietzsche once said, "Ethics are an invention of the weak to control the strong," and there was no way for the powerless Jewish woman to affirm her selfhood, satisfy her needs, and protect herself other than to try and manipulate the dominant male with guile and guilt. Of course, she had to hide her maneuvers beneath a barrage of self-deprecatory or sanctimonious words to avoid punishment. In addition to occupying a state of subservient, enforced dependency within the Jewish family, there were anti-semitic pressures from the dominant social group. So the Jewish woman (if not all Jews) had to become a good psychologist to survive oppression from within and from without the family. The Jewish Mother pattern, then, essentially developed to cope with people more dominant and powerful than herself. (And Lucy, of course, is stronger pound for pound than I am, and most of the time she cannot be made to do what she does not want to do by force alone—but this is getting ahead of the story.) The Jewish Mother personality pattern has persisted to this day, nourished by the Judaic-Christian tradition of American culture, even though it is a destructive pattern because guilt reduces self-esteem and blocks the expression of resentment.

Of course, one need not be Jewish to be a Jewish Mother; oppression and enforced dependency may produce the same pattern in Catholics or Baptists. And not all Jewish mothers are Jewish Mothers. My mother was a Jewish Mother, but not a Jewish Mother's Jewish Mother. (I am writing this not for the sake of public confession, but because it constitutes the background from which a critical observation of Lucy was made.)

My mother would utter such classical inanities as "Eat a little something, it's good for you," and I would feel I was too stupid to eat even if I had been hungry. Then later she might admonish me against getting fat or the necessity of avoiding indigestion by not overeating. When I was in grade school or graduate school she would say, "Work hard, make good grades—but don't be a bookworm, try to get along with the other kids." Then in graduate school when I was working hard and making good grades her attitude was, "Isn't it wonderful how successful your brother is!"

This is not written to be critical of her. She loved me in her own way as much as she could, and was unaware of the double-binds, inconsistencies, and demands. Indeed, I am grateful to her because I acquired my psychological sensitivity through interacting with her. The point here is that I was exposed to all the classic Jewish "double-binds," and I was influenced by them in ways which affected my behavior toward Lucy and Lucy's toward me.

Jewish Mother Guilt Dynamics

When Lucy was two years old we discovered that she was sensitive to and could be manipulated by Jewish Mother guilt dynamics. We first discovered this when attempting to change her diet.

During Lucy's infancy she was fed a baby formula by bottle. She had an immediate sucking reflex and nursed greedily and would take the formula well from Jane,

Steve, or me—whomever was holding her. Most of the time it was Jane, whom she preferred. At about one year we began to wean her to solid foods—Gerber's baby foods. Soon she was eating a wide variety of Gerber's prepared mush-like fruits, vegetables, meats, and cereals. She then began to get finicky and to eat only the cereals, avoiding the meat dishes. We felt she should have more meat since chimpanzees eat meat in nature and, except for a more highly evolved cerebral cortex and central nervous system, chimpanzees and human beings are physiologically pretty much the same. (Also, we knew from the autopsy on Charlie Brown that he had had a diet too low in protein and too rich in cholesterol, as even though he was only four when he died he was beginning to have fatty deposits in the arteries.) So Jane and I tried to get her to eat a diet developed by Dr. Herbert Ratcliffe of the Philadelphia Zoo. Actually, it was a modification of Dr. Ratcliffe's diet, in which Purina dog food was substituted for Dr. Ratcliffe's mixture of grains. It consisted of Purina Dog Chow, carrots, and beef organ meats, particularly liver, spleen, heart and pancreas, all ground up and cooked together with minerals and vitamins then added. Nutritionally, it was great. It had plenty of protein and carbohydrate and fats in the correct proportion, as well as vitamins, cod liver oil, and minerals, even trace minerals. In fact, it had everything in it Lucy needed. There was just one thing wrong with it. It tasted horrible and Lucy wouldn't eat it.

We knew we had to change her diet somehow because by then she had become such a picky eater that she would not even eat her cereal unless it was disguised by mixing it heavily with raspberry pudding. We first tried starving her, but at this endeavor we lost.

Lucy went two days without food. She cried, screamed, and tore her hair, but she would not eat. She would reach for the Ratcliffe, bring it near her mouth, then gag, sputter, and choke herself, but she would not eat it. She was a pitiful sight, running to the pantry or to the refrigerator door, pulling at the handle, screaming, kicking, and crying—and after two days of distress we gave in, having decided to try a more subtle approach. After all, we were human beings. We had a more highly developed cerebral cortex, a good education—all the advantages.

We gave her Gerber's cereal generously laced with raspberry pudding, her favorite flavor, but we surreptitiously mixed in a spoonful or two of Ratcliffe. The idea, of course, was to add a little more of the formula each day, gradually increasing the amount until Lucy would be eating the Ratcliffe diet. It was great theory, but it did not work. Lucy greedily ate the first few bites then seemed to taste the hidden Ratcliffe, because she looked suspiciously at Jane and me and would eat no more. At that point I felt frustrated and angry, but I was totally unable to express my anger at such a helpless, starving child. The Jewish Mother in me then emerged spontaneously.

Without thinking about it at all I pleaded, "For God's sake, Lucy, think of the starving chimps in Africa!" For several moments Lucy did not move. She looked at me with a strange expression I had never seen before. Then she slowly began to eat. After a bite or two I added, "Take at least three more bites for your poor suffering father who loves you." Her tempo increased immediately. She then acted as though "eating everything on her plate" was a matter of great moral virtue. We did not believe this at first. But to this day, almost eight years later, we have verified the observation countless times. If Lucy does not eat what we want her to eat or as much as we think she should, I need only remind her of the "starving chimps in Africa" (it was the Chinese in my childhood) or whine, "How could you do this to me?" and Lucy begins to eat greedily. We have demonstrated this to our friends many times.

Lucy clearly understands the general sense of my words, though she is not responding to the words alone. Jane, for example, can say the same words, invoke the same deity, moralize, even curse her. But if Lucy does not want to eat she refuses to do so. Then if Jane asks me for help, I can say the same words she has just uttered, and Lucy will eat immediately. Sometimes all that is necessary is for Jane to say "Maury?" with the proper helpless whine and Lucy, seeing me coming, looks guilty and begins to eat. This is particularly true at breakfast.

When Lucy gets up in the morning she has "coffee" with Jane and me. Her coffee consists of a large mug of warm milk with a spoonful of coffee added for flavor, and to make it look more like what we are drinking. Jane usually makes it, and Lucy often will only taste it before exhibiting elaborate disinterest. For example, she will taste it and then turn her head away, particularly if the temperature is not just right. When she does this I need only to look unhappy and ask with a whine, "Lucy, how could you do this to your mother?", and she starts making her good food sounds and drinking greedily.

During the past eight years Jane and I have often speculated as to why Lucy will eat so much better for me than for her. Jane says it is because I surround eating with more affect than she can summon. When Lucy will not eat what she has fixed for her Jane feels rejecting and usually could not care less whether, as she once put it, "the ungrateful little bitch eats or not." But I know better! It's because Jane is not Jewish.

I can remember when I was five or six and thought the whole world depended on whether or not I ate my spinach. When my mother urged "think of the starving Chinese," I ate quickly—so they would not starve, because if they did it would be all my fault. Wilhelm Reich once said, "The past is alive in the present in the form of character attitudes," and I believe that Lucy, since she is less sophisticated verbally than we are, responds to the whole person.

There may be other factors also which make Lucy susceptible to manipulations by guilt. It has been demonstrated that chimpanzees have a self-concept and can be both subject and object in their own perceptual fields. They can recognize themselves in a mirror, for example, and I believe are the only primate other than man who can do so. The fact that they can perceive themselves may make them vulnerable to negative evaluations of themselves and thus susceptible to manipulations by guilt. Lucy certainly responds to guilt-producing communications as though she has a primitive concept of right or wrong. I word it "right *or* wrong" because I think for her there are no shades of gray. We can see her creeping furtively through the house, not looking directly at us, and know she has hidden contraband: a cigarette lighter smuggled in her palm, a key hidden in her mouth, a stolen screwdriver for dismantling electrical fixtures. Or if Jane and I are talking and suddenly realize that Lucy is out of sight and silent, we know she is doing some forbidden thing: playing with something dangerous and forbidden we have failed to lock up such as household poisons, firearms, or matches. Or she is raiding the liquor cabinet or the refrigerator—lesser crimes, but still sins. Primitive concepts of good and bad are no real deterrent for her (when we are not present), any more than complexly verbalized sanctimony deters the more verbal primate, as the general public learned in the era of Nixonian morality.

Lucy rocks when frustrated and we respond with guilty feelings, particularly Jane. It is less effective with me. For example, she will sit still and rock back and forth, her movements regular and repetitive, looking neither to the right nor to the left, staring straight ahead. She at these times exhibits a kind of repetitive behavior often seen in autistic children and her rocking is much more routinized and stereotyped than when she is simply anxious. While this behavior may have a self-stimulation function, it elicits a quick, cuddling, comforting response from her mother, which disappears immediately once Jane is gone. We know this from surreptitiously viewing Lucy on closed-circuit televison. When Lucy is locked in her room she may protest and start to rock, but the rocking disappears when Jane leaves, only to return when Lucy hears Jane's footsteps approaching. But I am digressing. Right now I want to return to the issue of the Jewish Mother, as there is further evidence that Lucy perceives and responds to this form of interpersonal relatedness.

For almost eight years Lucy continued to eat better for me than for Jane. Even when Lucy was sick and had little appetite I could always get her to eat or to take her medicine by being a Jewish Mother with her. This differential capacity to influence Lucy always amused me and irritated Jane. It disgusted some of our friends to see a full-grown man acting like a Jewish Mother to control what was to them a beast of the forest, rather than a beloved, if somewhat different daughter. In any case, all of this ended dramatically in the summer of '74.

A Gestalt Workshop Changes Things

I had gone to the American Academy of Psychotherapists workshop in Claremont, California, on Wednesday, July the 3rd. That wonderful workshop lasted until Sunday, and I then traveled up the California coast to Esalen at Big Sur. There I attended a workshop led by George I. and Judith Brown. George and Judith had been trained in Gestalt therapy by Fritz Perls and in psychosynthesis by Roberto Assagioli, and they had combined individual Gestalt work and Psychosynthesis fantasy trips in a very effective manner. The workshop lasted two weeks and was called "Supervisory Workshop in Gestalt Therapy for Psychotherapists." I participated fully in the workshop and was in the "working chair" (the Browns prefer this term to "hot seat"), on an average of once each day. It was the most emotionally powerful two weeks of my life. Several times I lost control completely and exploded into a frenzy of rage, grief, tears, or joy. The content was usually unfinished business about some introjected but unintegrated fragment of an identification with my mother.

Although I had been involved in many forms of psychotherapy—as a patient, student, teacher, consultant, or practitioner—for about twenty-three years, some of the unfinished business was entirely new to my awareness. (I suspect that Freud was right for the wrong reasons in "Analysis Terminable and Interminable." It is likely that personality, the interpersonal world, and the interactions between ourselves and the world are so complex, that everyone always has some unfinished business.) In any event, I worked on my unintegrated but internalized Jewish Mother.

As part of the Gestalt work I "became" my mother with great histrionic, often hysterical, flourish and fervor. I raved and ranted, wore sackcloth and ashes, cursed God and beseeched His mercy. I went around the room putting each member of the workshop in as many Jewish double binds as I could recall: "Be a scholar, we are people of the Book; but don't be a bookworm, and get along with the people next door!" "Relax, take it easy, you don't have to be perfect; but be careful and don't make any mistakes!" "Be yourself, I love you just like you are; but it would be nice if you could make something out of yourself!" "Be a go-getter, be a success; but don't offend anybody and don't be aggressive, because Mother loves a nice boy!" And, of course, there was, "Show Mother that you love me—eat a little something!"

At the end of the two weeks I knew that the workshop had been a wonderfully powerful experience for me. I left Esalen with a warm positive glow which I hoped would continue. I felt better and liked myself better than I had in years. As a therapist I knew that cathartic and abreactive experiences can be very powerful at the moment they occur but may fade away if they are not integrated, and I knew I would have to wait and see if the changes I made in the workshop would be lasting ones.

When I returned home I was delighted to find a novel evidence of integration: Lucy wouldn't eat for me! She still loved me, of course. After an absence of three weeks she greeted me at the door with her warm love sounds and covered my mouth with hers in a chimpanzee kiss. But she would not eat for me. I told her that the chimpanzees were becoming an endangered species; that if she did not eat I would tell Jane Goodall on her; that by not eating she was hurting my feelings and ruining my life! But she still would not eat unless, of course, she damn well wanted to herself—and I now could care less.

She interacted with four Jewish Mothers: my own mother, two others who were actually Jewish, and one who was Catholic.

At two years Lucy was a sweet, lovable, tender, and charming bundle of contact comfort. She loved to kiss and cuddle, and seemed to find her greatest joys in closeness to her family, particularly her mother. She was into every-thing, constantly curious about her world, but exploring it only in graduated doses, never straying too far from her mother's side. When visitors came she loved their children, leading them by the hand throughout the house, showing them her toys, sharing bits of food and other goodies. With strange adults she seemed to have a characteristic warmth combined with curiosity, sitting next to them on the sofa, gently going through the women's purses, appropriating the hats or pipes of the men, amusing everyone by her somewhat parodied imitations of them as she studied her face in a compact mirror, smoked a pipe, or strutted about the living room with a hat or shawl on her head, covering her eyes so that she bumped into furniture, chuckling loudly. Emily Hahn, who visited almost all the chimp colonies in the United States while writing her book, *On the Side of the Apes*, was quite taken with Lucy's sweet and gentle nature. She was contemplating a second tour and wrote ". . . It was Lucy I thought of," as she planned her trip.

In spite of all her lovable qualities, even though she could eat at the same table, exhibiting manners and dexterity beyond those of a human child, as we know, she was not toilet-trained. (What greater sin could there be in a two-year-old Jewish child?) Also, she could not carry on the family name; she was different (what will the neighbors think?) and Lucy herself had few internal controls over her exploratory and self-assertive curiosity. Chimps, like human children, need clear external limits to control and define the permissibility of their own behavior. My parents with Lucy (as with me forty years earlier) were unable to set clear limits; my father because of his gentle, lovable passivity; my mother because of her mixture of love and aggression concretized as niceness. When Lucy would climb all over her, wrinkling her clothes, messing her hair, stealing her cosmetics, she could not say, "Goddammit, stop that!" She might feel that kind of hostility when her personhood was invaded, but she would communicate it to Lucy only through her tensed muscles and posture. She would say, "Isn't Lucy cute," and try and distract her by offering her a toy or suggesting an alternative behavior. Her indirect communication would leave the responsibility for changing her action entirely up to Lucy. "Wouldn't you like to go outside and play in a tree, like your daddy used to do when he was little?" she might say, forcing a laugh as Lucy stood on her shoulders digging in her hair. Lucy then would either ignore her mixed pleas or continue whatever she was doing with increased gusto. I always felt Lucy and my mother could not stand one another but that neither one of them would admit it. Lucy never did bite either one of my parents, but she was too aggressively playful for their tastes, and they gradually visited her less and less. Lucy did bite Rose, though, a Catholic Jewish Mother.

Rose was a social worker who joined a group I conducted in my home because she was experiencing difficulty with her husband and with her child, who was three years old at the time. It is the difficulty she was having with her child that is relevant here. "He's only three," she complained, "and I can't stop him from doing anything. He is into everything, and even climbs up on top of the house, and neither my husband nor I can stop him. I am afraid he's going to hurt himself because he doesn't know his own limitations. I've tried punishing him and he just ignores me."

As she described these difficulties with her child she was smiling, and seemed to be experiencing some sort of satisfaction. Her voice was often no louder than a whisper. Her smile would increase as she described some particularly aggressive feat, such as her son climbing on top of the house. She was a slender, attractive blond, who had basically a good figure even though her shoulders curved inward, giving her chest a slightly concave appearance which I have come to associate with aggression-blocked and depressed people.

She was intelligent, highly articulate, and sensitive, though she tended to be one of the more passive members of the group. Whenever she came to the group she could see Lucy in her roof-top room and she was fascinated with her. She was always asking questions about Lucy, and her face was more alive and animated when talking about Lucy than at any other time. She asked if she could visit us when the group was not meeting and play with Lucy. I had reservations about it, but I liked Rose and at that time Lucy had never bitten anyone. So I told Rose she was welcome.

The following week Rose appeared, ready to meet and play with Lucy. As I introduced them I had a momentary flash of anxiety as I noticed Rose's dress seemed inappropriate for playing with Lucy. She was dressed more like a Junior League fashion model than the blue-jean clad playmates Lucy usually had. But my anxiety was momentary. It disappeared completely when I saw Lucy's initial response to Rose. It was warm and loving. She kissed Rose on the mouth, put her arm about her neck, and sat beside her to search her handbag for goodies. Rose seemed relaxed, so I relaxed and left them together. Half an hour later Lucy bit Rose on the hand. The bite was bad enough to hurt and break the skin, but no stitches were required.

I asked Rose what had happened and at first she said she did not know. Lucy had been very playful, had opened her purse, taken her hairbrush and had brushed Rose's hair and her own. Then, Rose said, Lucy had invited her to play tickle-chase, and the game had been fun for both of them. They would tackle one another and take turns chasing one another throughout the living area of the house. The game had gone very well, Rose said, and she thought perhaps Lucy had bitten her accidentally. This was her only explanation at the time.

Rose finished my group, which was a month-long personal growth group, and she also was in psychotherapy with another therapist. I did not see her again for a year, then I met her in a restaurant. She seemed to have grown considerably, appearing much more animated and alive. She was still interested in Lucy. By then I had started writing this book, so I asked her if she ever had remembered any more than she had already told me about being bitten. "Oh, yes," she replied with an enthusiasm which would have been atypical of her a year before. "I understand it perfectly. As we played chase, Lucy became more and more aggressive, and I could set no limits for her. It was the same problem I had with my own child. Finally Lucy became very, very aggressive and had to set her own limits, which she did by biting me." I agreed with Rose's interpretation, as it was consistent both with the personality dynamics of Rose, as I understood them, and Lucy's behavior toward other Jewish Mothers. Actually, it was typical of Lucy's behavior toward anyone who could not say "No" with the emotion to demonstrate that "No" meant *"No!"* The bite had the function of ending the play, setting limits on aggression, and establishing a relationship of dominance in which the victim was unlikely to incite further aggressive play. The bites Lucy rendered at these times were always much milder than bites administered in rage. Often they did not even break the skin.

In a sense, Lucy had established a limit of Lucy, by Lucy and for Lucy, so she would not perish from the earth in an environment to which evolution had not adapted her.

CHAPTER THREE

What Will the Neighbors Think?

LIVING WITH Lucy as a member of the family created many complications for us socially. Lucy made many friends for us: zoologists, ethologists, and researchers of all sorts; university students from many places visited and were charmed by Lucy, as were many of our own students. But she had a negative effect on our pre-existing relationships.

Our friends and relatives began to visit us less frequently, for they had difficulty relating to Lucy. This was particularly true of people who were afraid of Lucy. They hesitated to say "I am afraid of your daughter," much less "Keep that Goddamn ape off me." It was easier to avoid us. When Lucy was little, friends or relatives would bring their young children over to visit her. Lucy was fascinated with very young children, and was always gentle with them if the child was smaller than she was.

I had been brought up to be a very conventional person, and I gained great satisfaction from breaking away from the teachings of my youth and living such an unconventional life. I always had a sense of humor about how Lucy affected our relationships with other people. When some people regarded us as kooks or mad scientists, it amused me. Also, I could be openly aggressive with Lucy in humorous ways. For example, the morning I wrote this page I awakened in my bedroom and walked across the house, saw Lucy and yelled, "Jane, Jane, there's a full-grown chimp loose in the house!" and Jane yelled back, "No, it's a monster!" and everybody laughed, Lucy included.

Cocktails

One time when Lucy was about three years old the then chairman of an academic department and his wife were having pre-dinner cocktails with us in the living room. The chairman's wife was a very formal lady dressed in exquisite taste, quite conscious of her husband's social position. She had so wanted to meet Lucy. "Adorable, just adorable," she cooed, shrinking from Lucy's outstretched hand when they first met. She was drinking a whiskey sour, sitting very proper and erect. Her husband was asking all the polite questions about Lucy, trying to look natural but having a hard time as Lucy was going to eat at the table with us. I saw Lucy eyeing the lady's drink greedily, but not soon enough. In a flash of black hairy motion she had snatched the drink from the hand of this startled but dignified lady, swallowed half of it, dug the cherry out with her fingers and handed the remainder back to our shaken guest. I think, however, this startled her less than my reaction to it. At that time I was preoccupied with maintaining control and dominance, and I felt I had to show Lucy that she must act properly if she was going to live in a world of human beings. So a fraction of a second after she had returned the glass, I grabbed her and bit her on the ear, the only place on her body both biteable and tender. Lucy screamed but apparently realized that she deserved it for she did not struggle or fight back. I was so shaken by the encounter that I failed to notice that the chairman's wife did not touch the remainder of her drink all evening, and I should have noticed because many people will not drink after Lucy has drunk out of the same glass. They never came to dinner again, nor did they invite us to their house.

We discovered that Lucy had a passion for alcohol when she was only three.

We had an orchard on the northwest side of our acreage. It was a small orchard, about thirty trees—plum, peach, cherry, and apple. In most years only the apples would produce a good crop because of the instability of the weather. We would have a warm and beautiful week or two in February or March. The sun would shine through the winter clouds and the temperature would climb into the seventies. The fruit trees, thinking it was summer, would blossom only to be frozen in an ice storm around the first of April. But the apple trees were late bloomers and we had enough varieties of dwarf apples that even if some did freeze enough would bloom late to make a good crop. We loved to take Lucy into the orchard to play. When she saw the trees laden with fruit, she would hoot her good food sounds—low, regular, gutteral hoots—and run joyfully into the trees. She would pick apples right and left, usually taking only a bite or two before discarding one in favor of another. Shortly, however, her behavior changed dramatically.

Instead of climbing into the tree or standing bipedally and picking fruit off the tree, we noticed that she would run underneath the tree and eat the rotten ones off the ground. I first experienced a feeling of disgust when I saw this, and then fear, thinking that Lucy might get sick as she was eating only very rotten apples. They were so overripe that they had lost their shape and were discolored, unappetizing masses of smelly pulp. Yet Lucy consistently refused the "good" apples I offered her, and continued to eat large quantities of rotten ones. Sometimes she would eat rotten apples until I was sure she would be sick, but she did not behave as though she were sick. She would lie on her back holding her stomach in a most uncharacteristic posture. Instead of seeming sick, she seemed happy, often laughing and giggling. Then Jane and I realized that she was getting a "high" on the natural fermentation as the apples rotted. We then started giving her a drink before dinner, or whenever Jane and I had one, and this behavior in the orchard discontinued. Since that time she has eaten ripe but not rotten apples. When she once visited us, I mentioned this experience to Jane Goodall, who had spent 12 years observing chimps in nature. Dr. Goodall said that she had not observed this behavior, which I joyfully interpreted to mean that Lucy was either a genius chimp, or that the enriched environment we had provided her had enabled her to actualize potentialities beyond those of chimpanzees in nature.

In some ways, Lucy is an ideal drinking companion. She is very appreciative, always making sounds of great delight when offered a drink. She never gets obnoxious, even when smashed to the brink of unconsciousness. Alcohol relaxes her and it improves her sense of humor, for she laughs and laughs, tickling herself, posturing before a mirror, making "crazy" faces and laughing at them.

There was another reason I enjoyed drinking with Lucy. I had felt guilt at offering Steve liquor when he was pubescent or even when he was a teenager. He was too young, it might hurt his liver and besides, what would the neighbors think? But I had none of these feelings with Lucy and she was so grateful that I felt appreciated. So each night before dinner we would fix Lucy a cocktail or two, a gin and tonic in the summer and a whiskey sour or a Jack Daniels and 7-Up in the winter. Lucy would then sit in the living room on a couch and drink them with us, sometimes staring out the window at the pond, sometimes lying on her back looking at the pictures in a magazine which she manipulated with her feet, or when really high, dancing about the room and playing games. Usually she had only one or two; three would leave her blind drunk.

Lucy developed a very discriminating taste in liquor. Jane and I preferred Gordon's or Beefeater's gin, and Jack

Daniels bourbon. As we noticed our monthly liquor bill we began to debate buying a cheaper gin and bourbon for Lucy. Also, during that period of our lives, we were drinking wine with dinner and we usually liked a fine Chablis or Rosé. Between the three of us, with each person having two or three drinks each evening and a glass or two of wine with dinner, it was amounting to a considerable monthly sum. So we decided to try and save money by buying a cheaper brand for Lucy.

Since Lucy loves fruit, and since chimpanzees in nature also eat large quantities of fruit enthusiastically, I bought a case of Boone's Farm apple wine and Boone's Farm strawberry wine to start Lucy on. We sat down to dinner with high hopes that our problem would be solved. Lucy drank her first glass of Boone's Farm apple with great relish, then she noticed that Jane and I were drinking something else. Actually it was a superb French Chablis, much more expensive than the seventy-eight cents per fifth we had paid for the Boone's Farm, even though it was roughly the same color. But apart from color, there was no resemblance. Lucy stopped, looked suspiciously at us, and did not move for several seconds. Then she reached across the table, took my glass and held it to her lips, looking at me as though I were a traitor. Slowly she sipped the Chablis. Her eyes widened, she made her good food sounds, and quickly drained the glass. Then she grabbed Jane's glass and the sequence was repeated. We immediately thought the wine situation was hopeless, but it proved not to be so, because of Lucy's imitative identification with us. She then took her glass of seventy-eight cents per fifth Boone's Farm apple and held it across the table to my lips. I drank it while Lucy held the glass and then I refilled the glass from the bottle. After seeing me drink it, Lucy drank a glass of Boone's Farm and has been fairly well satisfied with Boone's Farm since then.

I now keep the liquor and the mix locked in the pantry. But for several years I had an unlocked liquor cabinet in the kitchen, and Lucy would love to raid it. On rainy week-end days when Jane and I could not work outside, and when we did not want to leave Lucy in her room on the roof she would have the run of the house. Often I would be writing or working or taking a nap or talking to Jane, only to discover that Lucy had raided the liquor cabinet. Once she infuriated me by wasting a bottle of vodka.

At that time our living room had a concrete floor. So Lucy used the concrete floor as a mixing bowl. I had been in the kitchen talking to Jane and we noticed that Lucy was missing. Then we realized that neither of us had seen her for about thirty minutes. Going into the living room we found Lucy flat on her back so drunk she could not walk. She was lying on her back giggling. Every now and then she would try and stand up, get about half up on her legs, take a step, and then fall laughing to the floor. She had taken an unopened bottle of vodka, twisted off the cap, poured it out on the floor on top of little piles of Tang. She would take the Tang, make a little mound, saturate the mound with vodka, and then eat the Tang-vodka mix off the floor. I really believe that she mixed the Tang and the vodka more out of curiosity than an aesthetic preference for something like a screwdriver, because Lucy likes the taste of liquor in any form. As I pour a drink, for example, she will often put her face down next to the bottle and drink the straight liquor as it's being poured into the glass. The only liquor I have seen her refuse was straight creme de menthe.

If I mix Jane and me a drink and forget to mix Lucy one, she will often steal my drink, or Jane's. She will wait till we put it down between sips, show an elaborate disinterest in it by looking away and ostentatiously ignoring us, then suddenly grab it, scurry away, and drink it swiftly in the next room.

This is only one of the dozens of Lucy's expressions that are remarkably human.

A Foolish Consistency

As each man is in some ways like all other men, yet also is he unique. The same is true for Lucy. In some respects she is like other chimpanzees and yet she is uniquely Lucy. Her personality is remarkably consistent, so that her relationships with people are consistent. Even when she has not seen a person for a long time she still feels the same way about him when he returns. She does not show friendly affection toward a person on one day, and coldness or hostility toward him at other times, except when she is cycling. During this period, when her blood is charged with endocrines, she is more moody and inconsistent. Someone, doubtlessly a Freudian, defined the human female as "a creature who is psychotic once a month." When Lucy is in "full bloom," her genitals maximally swollen, she sometimes may exhibit a momentary impulsive hostility against someone toward whom she normally is warm and friendly. She might refuse an invitation from Steve to play with a brief but shrill scream at this time, for example. When she is not in estrus, however, she is quite consistent in her reactions toward people, apparently judging them by personality rather than by such superficial characteristics as how they dress. Similarly, people are remarkably consistent in their relationships with Lucy. The interaction between Lucy and other people has always seemed to us to reflect the character structure of both chimpanzee and human. It is interesting that Jane Goodall, in her observations of chimps in nature, found that they too had a stable individuality and interpersonal relationships. Here are some examples of the way Lucy and other people relate to each other.

Boris

Boris is a profoundly neurotic, very aggressive, manipulating man. Yet he often seems warm and friendly

because he sugar-coats his words with a greasy sanctimony, taking responsibility only for virtue. When things
go wrong he takes no responsibility and always blames the
victim. He once told me, "You can't win making any
public statement unless it's sanctimonious." He is rarely
behind his words and he is always manipulating other
people. Indeed, he feels acutely uncomfortable in any
situation or with any person he cannot control. He cannot
keep a relationship with another person going unless the
other person is dependent upon him. He is not open about
any of these dynamics, and may even be unaware of them
himself. Both to deceive himself and also to control others
he consistently creates complicated verbal situations to
which other people respond and in which they become
entrapped. For example, he will have people to dinner
who dislike one another and then play "peacemaker,"
confiding in each antagonist that the other "said in confi-
dence" that he wanted to be friendly. His relationships
with people become complicated, long-term networks of
mutual dependency for which he takes no responsibility.
He is aware that his life is devoid of love (he once called it
a disease), and that his relationships are filled with frustration for everyone around him, though he has no idea why
this is the case. As he once put it, "Everything I touch
turns to shit." Curiously, all this verbal defensiveness
disappears around Lucy. It is as though he senses that his
obsessional verbal defenses would be wasted on Lucy
(which they would be), and he thus becomes a much more
authentic person around her than around other people.
With Lucy he communicates forthrightly. Because of his
need to control, most of his communications are orders or
demands, and Lucy obeys him as his words and feelings
are correlated when he is giving orders, though at no other
time. Even though Boris often sounds like a drill sergeant
commanding troops, Lucy seems not to mind. Boris once
said, "I get along better with Lucy than with people." I am
sure that he has no idea why and that he will not recongize
himself, should he ever read this description. But we

eventually discouraged his visiting Lucy because we found it too unpleasant and destructive for us to be around him.

A person's character structure (in Wilhelm Reich's sense of consistent patterning of verbal and muscular armoring against the interpersonal world) is as clearly and consistently manifested in relationships with Lucy as in relationships with other people. The experience of Jean makes this clear.

Jean

Jean is a very attractive, sexually provocative, but sexually-repressed woman about thirty years old whom I met in a dramatic way. When Charlie Brown was three years old I was sitting behind the steering wheel of my car parallel parked in front of a grocery store. Charlie Brown was sitting on my lap, his hands on the steering wheel, avidly playing at driving the car as would a three-year-old human child. He would frequently turn the steering wheel, honk the horn, and was having a wonderful time. Carried away with joy he suddenly twisted around in my lap, threw his arms around my neck, and covered my mouth with his in a sustained kiss of loving gratitude. As we held the kiss—and I must say we rather than Charlie, as I had paternal pleasure from being kissed by such a charming and beloved son—at that precise moment a woman walked between my car and hers, parked three feet away. She saw the entire incident, and I could observe her face over Charlie's head, which was twisted sideways to permit a fuller coverage of my mouth with his. The woman stood, transfixed. Her face whitened. She must have stood there for fifteen to twenty seconds, her face a mask of unbelieving horror, before she continued into the grocery store shaking her head. Although I did not know the woman and the whole incident had not lasted a full minute, I was worried about an accusation of sexual perversion. This was a small southern-midwestern town before the so-called sexual revolution. The campus disturbances at Berkeley had not occurred. "Counter culture" was an abstract sociological concept, not a reality. And even today, many locals feel that Alex Comfort's marvelous book, *The Joy of Sex*, is a Communist plot to undermine the moral fiber of America. So when Steve returned from the grocery store I asked him to babysit with Charlie while I entered the store to explain to the woman the research aspects of having a chimpanzee in the car.

As I searched through the store for her, my mind filled with anxiety about the gaucheries I might commit. I could hardly say, for instance, "Excuse me, lady, I am not a sexual pervert, I was only kissing that chimpanzee for science." Nor could I utter the absurd thought that came to me, "It was all Charlie's fault!" Silly as all this sounds, though, I felt I must do something. I did not know the woman's name, though I had seen her on the campus, and I was concerned about the possibility of bad public relations. By then we had met in front of the coffee shelf between the aisles.

"Excuse me," I began. "My name is Dr. Temerlin. I want to explain about the research project, if you've got just a minute."

"Yes?" she replied, with a forced smile.

"I work for the psychology department, and the baby chimpanzee is part of a research . . ."

"I'm sorry," she interrupted. "Why are you telling me this? I've got some shopping to do and I'm not interested.

"I don't want you to misunderstand what you saw. When that chimp kissed me," I pleaded in desperation, "I—"

"Are you crazy?" she interrupted, confirming what by then was obvious. She had repressed the entire experience and thought that I *was* crazy. The idea of sexual contact between a man and a chimpanzee on the streets of a small town was so repugnant to her that within a few moments she had excluded the perception from her awareness. I was shocked at her repression of a harmless experience that had been so meaningful to me. Suddenly humor conquered my anxiety. I laughed aloud at the thought that under the laws of the state to be technically precise since Charlie Brown was a *male* chimp I was guilty of homosexual bestiality.

A few days later I happened to see her on the campus. She was an assistant professor in one of the language departments and I invited her to have a cup of coffee with me in the Student Union. After a more relaxed conversation about the university administration, the federal government, the Spanish department, and the like, I approached the subject of chimpanzees. As the conversation continued in this relaxed vein, I changed from the mad stranger who had spoken gibberish in the grocery store to a friend. Suddenly she said, "My God—I know what you were talking about in the store." She then discussed the whole incident with me. She realized that she had repressed the experience seconds after she had seen it.

We saw this same inability to perceive the novel many times. For example, it is a very common experience to stop at a traffic light with Lucy sitting on the front seat of the car and see people in adjoining cars look right at us, and not see Lucy. If this other car has children in it, though, they notice Lucy, as their perceptions have not yet been limited by cultural conditioning to perceive only the socially acceptable forms of "reality."

George

Once I attended a meeting in a federal prison. I was leading a group there, as were several other psychologists, in a program to try and reduce tensions among the staff and prisoners by using group therapy and encounter techniques. It was about fifty miles away from my home and there was snow and ice on the ground. My car broke down and I hitchhiked a ride home with another psychologist—a rather relaxed, groupy type, very competent as a therapist. He had an unusually good sense of humor, which he needed on this occasion.

We had to travel very slowly because the roads were so icy. We skidded back and forth many times. We had no tire chains and were not accustomed to driving in such conditions, which were unusually severe. All of our attention was on the driving. When we got home several hours later it was with great relief. Since my benefactor had another thirty minutes to drive to his home before him, he asked if he might use the bathroom first.

"Of course," I replied, unlocking the door. My thoughts were on our narrow escape on the icy roads, and I showed him where the bathroom was without a thought that Lucy might be loose in the house. George entered the bathroom, closed the door behind him, and I returned to the living room. Moments later I heard a strange sound, something between a gasp of astonishment and a scream of terror, and I realized that Lucy had been loose as I rushed back to the bathroom.

As George stood before the toilet, eyes closed, urinating with great relief, Lucy had quietly opened the door, entered the room, walked around him and caught the stream of urine in her open mouth. George had opened his eyes to find himself urinating into a grinning cavernous mouth, her huge canine teeth only an inch from his penis. He had been unable to suppress the quite unusual sound which spontaneously emerged. He did, however, make a quick recovery. As he told me what had happened his laugh was forced though his humor was genuine. Lucy, meanwhile, had gone into the living room happily gargling her tasty prize.

This interest of Lucy's in urination (male or female) is strange and fascinating. Since she was three-and-a-half or four years old, I have found it almost impossible to urinate in her presence without her attempting to catch the stream of urine in her mouth. Jane and I would be walking through the woods of our cattle ranch, for instance, and I would have to urinate. Lucy, though she was forty to fifty feet away, would often scramble up and down the bank of a river or rush through the trees to try and catch the stream of urine. I do not know why, but this behavior has always inhibited me. My head tells me that it cannot hurt her. I know that in chimp colonies mothers will sometimes hold their baby chimps aloft when they urinate to catch the stream of urine in their mouths. Further, I know that urine would not contain pathogenic bacteria, as long as the urinator is healthy. Nonetheless, I found myself inhibited and I have always stopped the stream of urine or turned around so that Lucy could not catch it in her mouth. Once I resolved that the remnants of the Judaic-Christian tradition in me, which says that the body's excretory products are dirty, was not going to stop me. I said that the next time Lucy tried this I was not going to inhibit myself, and I was going to urinate in her mouth. When Lucy next tried to catch the stream, I tried to relax my urinary sphincter and continue urinating. I was not successful at this; I always had an involuntary sphincter contraction which, I suppose, is a tribute to the persistence of inhibitions acquired during the early years of life.

The Meek Shall Inherit the Earth

We once had a married couple with whom we were quite close friends. We socialized with them on the average once a week, enjoyed their company, and they enjoyed us. However, they were an extremely conventional couple, each of whom had needs to be "nice," and they clearly could not take Lucy. She would jump on them, throw her arms around them and kiss them, grab their drinks and hassle them in other ways as well. She would steal the purse which the woman carried, open it and take out cosmetics and mirrors to play with. She would crawl into her lap with a "dirty" bottom, contaminating her dress with chimp feces. They could not say "No" and make it stick. They could not tolerate the aggression needed to lay the law down for Lucy and enforce it. Jane and I could, of course, but to do so we had to be extremely aggressive with Lucy, and our aggression offended them, as they felt we were being unnecessarily violent. Eventually, they faded away, and though we see one another occasionally, in the supermarket or on the campus, we no longer are close.

When Lucy was little we used to take her around town in the car with us. Now that she is full-grown, we don't do this very often as we are afraid of an incident. But when she was little Jane and I used to get a lot of pleasure taking her with us wherever we could. All kinds of interesting incidents occurred. On one occasion I was double-parked with Lucy in front of a T.G.&Y. store on Main Street while Steve went in the store to make a purchase. An adolescent walked by, the window was rolled down, and he saw Lucy sitting beside me. "Is that a monkey?" he asked. (I have always been offended when people refer to Lucy as a monkey rather than a chimpanzee. A monkey is lower on the evolutionary scale and does not have the self-awareness a chimpanzee does.) "What's he doing there?" the kid asked, though I had not answered his first question. "He's waiting for his brother," I said, "who's gone into the store." The young boy stared at me, clearly couldn't come up with an answer and finally just walked away.

When the movie *Planet of the Apes* was showing at a drive-in near our home, Jane and I took Lucy with us to see the movie. The cashier, also a young boy, looked at Lucy sitting between us, paused for a moment, and then made the decision himself. He handed us two tickets rather than three. I paid for them and said, "Thank you for letting her in free."

Lucy paid no attention to the movie whatsoever, nor have we seen her express an interest in chimpanzees when she sees them on television. Lucy's lack of interest in television is interesting. She shows no interest in it whatsoever, though she will turn it off if we are watching it and she wants attention from us. Or when she slept with us and we were watching a late movie and she could not sleep, she would frequently jump up and turn it off. I believe this must be an individual characteristic of Lucy, because Emily Hahn, in her writings on apes, quotes a Dr. G. H. Bourne as saying the chimps in his Yerkes colony watch TV, which he uses to reduce "cage fatigue." I prefer to think that Lucy, having had all the cultural advantages since birth, has too much taste to enjoy TV, and is too happy to need escape.

Housekeepers and Babysitters

During the past ten years we were always short of babysitters and housekeepers. Even when Lucy was very little we had difficulty finding babysitters whom we would trust who would take care of her for more than one evening. Usually she had to be asleep before Jane and I felt it was safe to go out. The problem with housekeepers was even worse. We needed a housekeeper constantly because of the toilet training problem, and also because she kept the house in constant disarray. But we never had a house-keeper who stayed for very long. They always quit for one reason or another, even when we did everything possible to protect them from Lucy. Two of them stand out in my mind with great and painful clarity.

Shelly

We hired Shelly as a housekeeper even though we were ambivalent about doing so for two reasons. First, he was a male and we had never tried a male housekeeper before. I felt that housekeeping was a job beneath the dignity of the male, that it was a female job, illustrating of course a residual of a sexist bias acquired early in life. But even more objectionable was Shelly's compulsive politenesss. If I were to describe him clinically in one non-technical phrase, it would be as a "nice Jewish boy." Shelly was very obviously Jewish and he was very obviously nice. Even after he had worked there for a month, he would have to start each afternoon's session (he worked half-time, going to school in the mornings) by asking me, "How is your wife?" then, "How is your mother?" and the same about my father. He had to get a report on my family's health before he could start to work. He smiled constantly, even when there was nothing to smile about as far as I could see. He was a very gentle man, a "nebbish" in the Jewish argot which he frequently used. We were sure that Lucy would be too much for him, and I initially told him "no" when he applied for the job. Shelly said, though, that he always got along well with both animals and with people, that he was an excellent housekeeper, that he enjoyed housekeeping, and that it was his only way to get through college. We were desparate for a housekeeper at the time, there were no other applicants, so we decided to give Shelly a try.

We introduced Shelly to Lucy as gently as we could by asking him to sit still on the couch, not to move, and not to talk, but to let Lucy come to him and let her write the script. Lucy charged into the room like a Mexican fighting bull into the arena, noticed Shelly, jumped into his lap, and covered his mouth and half his nose with her mouth. Even though we had coached Shelly that this would happen and to sit still and just take it (enjoying it if he could), Shelly recoiled. His face whitened and he shrank back into the chair and, of course, into himself. While Lucy did not bite him at that time, I believe that she developed a negative attitude toward him. After all, her intentions were pure. She was greeting him in what to her was a warm and cordial fashion, and he shrank away from her. She could only have interpreted his behavior as a form of rejection. It is interesting that Shelly, when asked how he felt about Lucy immediately after she kissed him, said, "I was not afraid." ("There are no negatives in the unconscious." S. Freud) Since at that moment he looked like a cross between a melted marshmellow and Casper Milquetoast, I was not convinced. At the same time I know how unaware people can be of their own feelings, so I decided to let the matter go.

Lucy went about her business and Shelly went about his. Shelly was a fantastically verbal housekeeper. No matter what he was doing, he was talking; to himself, if no one else was present, or to Jane or me if we were in the same room. I had told Shelly not to talk to me while he was working as I had other things to do, but he was one of these people who uses talking to reduce the intensity of his own experience. I doubt that he was aware of what he was saying most of the time. Certainly he was not listening while he was talking.

We had told Shelly always to knock before he entered a room to see if Lucy was out and if something was going on with her which he should not interrupt. One afternoon Shelly knocked on the door, talking constantly, and Jane called, "Don't come in, Lucy is loose." Shelly immediately opened the door, entered the room, walked directly to the freezer, opened the freezer door preparing to thaw out frozen food for dinner and Lucy bit him. It was not a bad bite. Lucy bit him gently, just enough to tell him that she did not like his messing around in her kitchen with her food—a rule she had made herself. We had told Shelly, who still was talking constantly, not to do housework such as scrub floors, open the refrigerator, or move furniture when Lucy was in the room. And, of course, after that incident we told him again. Shelly handled the bite well. He petted Lucy, he did not yell but he also did not learn. Lucy went into the other room and went about her business and Shelly continued his work in the kitchen.

A week or so later Lucy was sitting on the sofa looking through a comic book. This was Shelly's day to scrub the floors, and to scrub the floors and wax them he had to move the couch. Talking all the while, unaware of what was going on, with his mind only on his own inner life (if on anything), he started to move the furniture. He moved three chairs and all went well, but then he decided to move the couch that Lucy was sitting on. In order not to disturb Lucy he didn't want to ask her to get down from the couch and sit elsewhere, so he lifted up the end of the couch with Lucy on it. It happened so fast that he did not even see himself get bit. He looked down to find his hand bloody. His actions were in direct violation of Lucy's sense of territoriality. It was her living room, her house, and her furniture, and she definitely regarded Shelly as an intruder. We had known this, of course, and warned him repeatedly. Anyway, at that point we decided to check our liability insurance to be sure that it was still in force and to let Shelly go as a housekeeper. I was glad that it was not necessary to fire Shelly, however, as he called in the next day to say that he would have to take a job which paid him more money as his wife needed an operation.

Applying a clinical interpretation to the affair of Shelly, which is consistent with Lucy's behavior in the past ten years, I would say that Shelly was unaware of his own aggression, thus he was unaware of the messages he was sending which incited Lucy into an aggressive act; Lucy was unimpressed with his obsessive politeness. Generally speaking, the more passive the person with whom Lucy interacts, the more aggressive she is likely to be.

The best housekeeper we ever had was Mrs. Glup. We called her Gluppy. She had raised five kids, was about sixty, and would take no nonsense from anybody. She could put out warmth, she liked Lucy, and liked children. At the same time, she'd had enough of them crawl on top of the furniture or all over her to be able to say, "No—goddammit, go to your room and clean it up." She and Lucy got along very well. Unfortunately, she did not stay with us very long because she had to be cleaning house and taking care of somebody. She was the perfect housekeeper for us, but her relatives needed her more and she moved in with one of her daughters to manage her home and take care of the grandchild while her daughter worked.

Maury and Jane on the ranch. Lucy is climbing in the tree. Note swollen genitals indicating that she is in estrus.

CHAPTER FOUR

Picnic

A FEW YEARS ago we bought a cattle ranch, 758 acres, on which Lucy could roam free much like a chimpanzee in Tanzania. There were hills and valleys, small lakes, and creeks, and plenty of dense forest. The forest was of blackjack oak, live oak, and hickory trees, with occasional groves of pecan, persimmon, wild plum, and grape. So Lucy had her choice of the kind of land on which she wanted to climb, run, and explore.

Spending a day with Lucy in the forests of our ranch were some of our happiest times. All we had to do was say the word "ranch" and Lucy would become extremely excited. She would watch from her roof-top room as we prepared the pickup. She would scream with excitement, jump up and down, shake the wire wall of her room as she saw us make preparations for the trip. I would put a power saw and a two-gallon can of gasoline in the back of the truck, as I often cut firewood while Jane and Lucy played. We would get a thermos jug of water, a sack of fruit, and a sandwich for Jane and me, and sometimes a beer, put them in the truck and then open Lucy's door. She would make a beeline through the house, sometimes grabbing her leash on the way, run out the front door to the truck, and hop in. It was her favorite time.

She would eat contentedly between us in the pickup, usually eating a nectarine or a peach. We'd turn east a mile south of our house. The road was smooth and Lucy always relaxed, watching the scenery. Occasionally we would pass horses. Sometimes people would be riding horses along the highway or in front of their acreages, and Lucy would "boo" them—a sound she usually utters when encountering animals larger than herself. She loves this ride, carefully watching the other cars and the trees and the scenery—the trees turning brown and golden in the autumn and richly green in the summer. We were interested in the dead animals on the road, feeling they would say something about the kind and density of animals on our preserve. Most of the time they were dead armadillos. When we first bought the ranch three years earlier, they were almost never seen. Now it's not unusual to pass three or four each trip. The armadillos clearly have been migrating. Sometimes, too, we'll pass a dead racoon or coyote, though oppossums and skunks are much more common. On our ranch we have seen wild turkey, rabbit, deer, squirrels, possum, skunk, coyote, armadillos, and half a dozen species of snakes.

Lucy's Cat

Lucy is interested in the dead animals, too, particularly since her cat died. That's a story in itself. One time we had gotten her a kitten. Her first response to the kitten was hostile; she tried to kill it. After several hostile introductions, the kitten suddenly started following Lucy and Lucy's attitude immediately changed. From that moment it became Lucy's cat and she always wanted it with her. The kitten would follow Lucy about, and Lucy would pick it up and carry it on her back much as a chimpanzee mother carries her infant offspring. She never wanted to be without her kitten. Sometimes Jane and I had to separate them so the kitten could eat because Lucy hated to part from her even for a few moments. She maintained a constant skin contact with the kitten. She would often pet and stroke the kitten while it was eating or drinking its milk. Lucy tasted the cat food, but did not like it, much as human children identify with their pets and want to be treated the same way, eating out of the pet bowl and wanting to sleep together. Sometimes Lucy would be too rough with the cat—particularly if there was a difference of opinion. For instance, if the cat did not want to go upstairs with Lucy, Lucy might rely on her strength, just grab the cat by the neck, hold her aloft and take her wherever whe wanted to go. They had a beautiful relationship for almost seven months, and Lucy was deeply attached to her cat. Unfortunately, though, one day the cat died. I was in the courtyard at the time and I heard a scream coming from inside Lucy's roof-top room. It was a different kind of scream from any I had ever heard and I rushed immediately to the roof of the house. There the cat was dead on the floor, of undetermined causes. Lucy was at the other end of the room, obviously quite shaken. She was staring at the body very intently, not moving a muscle. Lucy approached the body, reaching a forefinger towards it as though to poke it. But she never touched the body. She withdrew her hand rapidly and shook it, a movement indicating anxiety, just before making contact with the body. She clearly had some sort of understanding that the cat was dead, never to return, for she never looked for it again nor seemed to expect to play with it.

Three months after her cat died, Lucy was leafing through an issue of *Psychology Today*. She was turning the pages rapidly and casually as she usually does. This issue had an article on chimps and included a picture of Lucy and her cat. When Lucy came to the picture of herself and her cat, she stopped turning pages, and sat transfixed, staring at the picture. She stared for about five minutes without moving, and then starting signing in ASL "Lucy's cat, Lucy's cat, Lucy's cat—" altogether she signed "Lucy's cat" repeatedly for another fifteen minutes as she continued to stare at the picture. Her mood was one of thoughtful sadness.

Photograph courtesy Sue Savage.

Photograph courtesy Sue Savage.

At left, notice how Lucy's face shows love and concern for her cat. Above, Lucy still shows concern for Lucy's Cat, and the cat appears to feel quite at home on her friend's back.

Anyway, it was after her cat died that Lucy started showing interest in the dead animals on the road as we drove to the ranch on the weekends. She would stare at them and sometimes turn her head to watch the body as long as she could as the car passed. Once she saw a dead cat ahead in the road, and she became very excited, uttering a "boo" sound, and put her hand on the steering wheel and steered the car into the other lane around the body so the car would not run over it. To my knowledge Lucy had no further contact with cats for about two years. We decided not to get her another one. Then Steve, who was living in town, visited us with his cat. Lucy attacked savagely the minute she saw it and it took all Steve's strength and dominance to save the cat. Lucy clearly was trying to kill it. She screamed and charged the moment she saw it, charging in a direct line which knocked Steve out of the way, grabbed the cat and threw it high in the air. It would have been killed had Steve not intervened. A few weeks later Steve brought the same cat back. This may have been poor judgment, but the cat was sick and Steve could not medicate it properly in his apartment. Again Lucy screamed in fury and charged, again knocking Steve aside to get at the cat. I thought she was going to attack Steve she was so enraged when she grabbed the cat and pushed him away. Steve finally grabbed the cat at great risk, I felt, to himself and pushed Lucy away. Lucy then whimpered and looked uncomfortable. I do not understand this behavior, but it clearly had not happened before. She had never attacked any animal so savagely before her own cat had died, but is this so surprising? I have often seen aggression or hostility mobilized in humans as a response to grief and loss.

I have always been amazed at the behavioral psychologists who for years thought that man was the only animal that had consciousness and foresight. Traveling with Lucy in the front seat of the car illustrates the absurdity of that idea. She carefully watches the road, and if we are going too fast, she gets nervous and starts to rock. Occasionally, of course, we'll have near-misses; a car in the other lane will crowd us off or a passing car comes too close. Often Lucy can see these situations coming, and she will nervously whimper and start to rock. The clearest example, though, is that she hates to go over bridges, as they scare her. She sees them coming, sometimes as much as two or three hundred yards in the distance, and starts to rock, and sometimes to whimper as well.

There is a bridge about two miles north of our ranch about one hundred yards long, a rickety old bridge made of wooden cross-timbers imbedded in a metal frame. It makes a hideous noise as we drive over it, and sometimes *we* get scared driving across it. When Lucy sees it coming for as far as a half-mile away she begins to exhibit all the signs of anxiety—she whimpers, rocks, and occasionally grabs the hand of whoever is driving. On several occasions when we crossed the bridge too fast, and made too much noise, she went out of control and grabbed the steering wheel and tried to wrestle it from me.

Other than when she is scared and reaching out for contact comfort, Lucy does not like to be touched in the car. If we put our arm around her in a gesture of affection, she will move it and lean away. I suspect this is because she doesn't want to be distracted, much in the way that the driver of a car in a rainstorm doesn't want his lover to bother him—however much he loves her. We slow down for a small town: one filling station, two churches, two or three dilapidated houses needing a coat of paint, and Las Tapatias Mexican Restaurant, which is a converted second- or third-hand mobile home. We always have fantasies of walking in there for a Mexican dinner with Lucy, but we've never had the guts to do it. I mean this literally, too, since I have trouble these days digesting Mexican food. A few miles east of the restaurant we turn south on a rough country road. It has more chuckholes, and Lucy rocks more. She seems to see the chuckholes coming, and frequently braces herself before we hit a rough one.

We're in a much more rural country than where we live and the number of pickup trucks on the road increases. The standard fare is a gun rack in the back of the truck, holding a 22-rifle or a shotgun, the driver in a straw hat, and the car adorned with religious bumber stickers. "Wave if you love Jesus," or "Smile because God loves you." Occasionally a bumper sticker enlightens us that "Goat-ropers need love, too," or that "It's great to be a Christian." A mile north of the ranch we turn west on a dirt road. It's another half mile from the turn to the entrance to our ranch, and Lucy usually gets very excited at this turn, as though she knows fun time is coming.

The Ranch

The minute we get to the ranch the first thing Lucy does is jump out of the truck, walk two or three feet away and squat to urinate or defecate. It has always amazed me because it indicates she can control her bladder and bowel. When we get to the ranch, we open the barbed wire gate and usually drive deep into the ranch so we are completely hidden in groves of trees. This makes it better play country for Lucy, and easier for me to cut trees for firewood.

Lucy prefers to play here, too, as there are lots of leaves on the ground. There are no grass-burrs and stickers which plague her in open country, and here she can climb one tree after another. She always climbs a tree, plays in it for a few minutes, comes down and runs to another and climbs it. I have never seen her jump from one tree to another, as Tarzan and his chimpanzee did in the movies. The only exception to this is when she is descending a large ravine or canyon. If there are trees growing out of the sides, she may jump from one tree to another to get to the bottom in a hurry. But as we travel through flat country with oak, hickory, and pecan trees close together, she rapidly climbs one and descends it and walks along the ground on all fours to another before she climbs that one.

In the summer when the leaves fill the trees it's a lot of fun to play hide-and-go-seek with Lucy. When she was small she always stayed close to Jane, only gradually increasing the distance she would put between herself and Jane. We have seen her get too far away without realizing it, then scream in terror and run back. She was five years old before she voluntarily would let herself get out of sight. But now, as an adolescent, she initiates the hide-and-go-seek games with Steve, Jane, or me. At other times we start it, but she understands its fun aspects. She can be gone from sight in seconds, and it is all but impossible to find her. It *is* impossible to find her if she doesn't want to be found. And it's amazing that even though Lucy's color is predominantly black and the predominant color among the trees is green, somehow she is perfectly camouflaged. She can disappear while we are watching her, then seconds later she may reappear on another side of us. This, of course, is in dense growth.

The best way to find Lucy is not to look for her but to sit still with eyes open, moving the whole head rather than just the eyes, and to let visual impressions come in to one rather than looking outside for some clue to her location. Sometimes this works. If I just relax visually and let impressions come into me rather than staring or scanning, I may then notice an eye or a speck of black hair or a movement and I'll have found her; or expressed better, she will have been revealed to me, only I will lose her moments later. So I am not surprised that Jane Goodall spent six months in chimp country in Tanzania before she saw her first chimp, though she studied the woods each day with field glasses.

Lucy particularly loves to play in hidden clearings surrounded by trees. The clearings give her a chance to run at full speed for a moment, and then to disappear into dense brush. If we are in the middle of the clearing, she will run across it right at us, then just barely miss us and disappear into the woods again. Sometimes we will separate to try and see who can find Lucy first, Jane, Steve, or me. When we first started these games we were afraid of losing Lucy in the forest, as the ranch is about a mile and a half square, and Lucy runs so fast and ranges so far. But we discovered that no matter where she is, there is one way to bring her back, which is to act as though Jane or I are hurt. The best way to do this is to utter a high-pitched scream of distress. This can be heard a long way off and it never fails to bring Lucy back. There is no other way to get her to come back to the car when it is time to go, if she does not want to.

We have never been able to catch Lucy unawares in the forest. I have sat beneath the tree for hours watching her, or more properly watching for her, then tried to sneak up on her when I saw her in the distance. I never got more than fifty feet from her before she would notice me and react in some way.

It is interesting that we have never seen Lucy make a nest in the woods. Chimps in nature build nests, bending branches and twigs down, stomping them, until a roughly circular semi-flat nest is formed. Lucy builds nests in the living room of blankets and pillows, but we have never seen her do so in the woods. This could be an effect of her being raised as a human, or it could be because we have never spent the night in the open woods with her. However, that is unlikely because she builds nests in the house in the day time and plays in them. Thus, I think her lack of nest-building in the woods is a result of her experience with us. Probably a chimp in nature learns nest building by observing another chimp do it. Lucy has seen us arrange pillows and furniture, but not limbs and leaves.

When we are in the woods I usually wear a 22 pistol on my belt because we once had an epidemic of rabid skunks. Usually a nocturnal animal, the sick skunks would be wandering about in the daytime, and I was afraid one would bite Lucy, as she always chased small animals when she met them. Steve often carried a 22 rifle. Lucy was terrified of firearms. Steve and I liked to shoot at tin cans, but we could never do it with Lucy along. The one time we tried it she screamed in terror and ran into the woods, and it took us quite a while—I would say an hour anyway—to assure her that everything was all right and to come back.

The ranch has several persimmon groves. Lucy loves to eat wild persimmons. She starts making her good food sounds when she sees them in the distance, then runs toward them. She eats both ripe and green fruit, with none of the "puckering" of her mouth that green persimmons cause in humans. It is very interesting that neither Jane nor I have been able to get Lucy to share wild persimmons with us. She will share food at home. But even though the trees are covered with fruit, she will not hand us one when eating the wild persimmons.

There are about 100 head of cattle roaming about on the ranch, and when Lucy encounters them she "boos"—the low, soft "boo" sound she always makes when encountering an animal larger than she is. She keeps a respectful distance from them and runs screaming up a tree if one should suddenly run toward her. Yet this "cowardice or caution," whatever it is, disappears entirely when the cattle are caught in a corral or in a chute. Then she is very brave, even cruel. She runs on the top rail of the corral, rapidly on two feet around the tightly-caught herd, brandishing a stick or a tree limb like some prehistoric herdsman. She becomes very excited then, sometimes uttering loud whoops, and throwing her branch at the cattle. If one is caught in the squeeze chute for vaccinating or branding, she may hit the terrified creature on the back with a stick then run rapidly away.

Having animals larger than herself trapped and helpless is more than her under-evolved controls can bear, and she delights in the expression of an undisciplined aggression, like drunken cowboys terrorizing a town in a western movie. We cannot work the cattle at all, cannot brand or vaccinate or sort them when she is around, for her aggression makes them unmanageable even in the small confines of the corral. And once when Jane and two friends, Jack and Mary, were working cattle—I was not present at the time—Lucy got so carried away with aggression and excitement she bit Mary, the lowest ranking member of the human species present at the time, even though she likes Mary and gets along very well with her at other times. So when the cattle have to be worked, Lucy now stays home.

We loved to take Lucy into the forest on our ranch so much that we usually wanted to share this experience with people we liked. However, this proved to be a dangerous and depressing experience because Lucy is more aggressive in the forest. Twice she bit a friend. After Lucy bit Mary, we really should have known better, for the chances of Lucy biting someone are always greater with us present and when aggression is mobilized by some other factor such as the excitement of running free in the forest. Our presence seems to mobilize all kinds of protective, aggressive, or dominance responses which are less likely to occur when she is with people with whom she is less involved. But our hearts triumphed over our heads. The friend was Natasha Mann, and it happened like this: Natasha was a psychotherapist who had developed her own combination of movement and Gestalt therapy. I had met her in July of 1974 at the Claremont Conference of the American Academy of Psychotherapists, and liked her immediately. After that conference we traveled from Los Angeles to Esalen together, she to spend a week with Al Huang doing Tai Chi and visiting old friends, since she had worked there before, while I attended a workshop in Gestalt Therapy led by George I. and Judith Brown. We hit it off together and in a very short time I felt I had known Natasha all my life. I asked her to visit me and to conduct a movement and Gestalt workshop in my home. She was on her way to do groups in Europe at the time and was delighted to do so. She very much wanted ". . . to see Lucy and to study Lucy's movements."

I picked her up at the airport late one night and we entered the house about midnight. Lucy was asleep on the sofa in the living room and Jane was in the kitchen. Lucy slowly came out of her sleep, noticed Natasha, and liked her immediately. She jumped up, threw her arms around Natasha's neck, her legs around her waist, and covered Natasha's mouth with her own. I was relieved and delighted, saying, "You're in!" The next morning the sequence was repeated. Lucy seemed extremely fond of Natasha, and Natasha was enchanted with her. It was impressive to see how this wonderful woman, a specialist in dance and movement as well as psychotherapy, communicated nonverbally with Lucy. She seemed to be able to read her moods, and unlike many people, she sent no nonverbal messages to confuse or excite Lucy. Since her workshop did not begin until late in the afternoon, we decided to go into the forest with Lucy and walk for an hour or so. We got a sack of fruit and a thermos bottle of water and piled into the car, Lucy eagerly running before us to open the car door and hop in. Lucy and Jane rode in the back seat and Natasha and I sat in the front for the thirty-mile drive to the ranch. During the drive Lucy seemed more comfortable than usual. She ate an orange and a nectarine and rocked anxiously only when we were crossing the rickety wooden bridge. On the way down Natasha asked a question most people ask sooner or later, "What's Lucy's intelligence?" With Natasha I felt freer to answer naturally. "It depends on where you measure it—she's mentally retarded in the classroom and brighter than Einstein in the woods."

We entered the ranch on the south half, which meant that we had about 300 acres of woods to explore. I had to carry Lucy on my back the first hundred yards. It was fall and sticker weeds had gone to seed and Lucy was getting so many stickers in her feet as she walked across an open field towards the first woods that she refused to walk. As soon as we got to the edge of the woods, Lucy jumped down off my back, and Jane, Natasha, and I began following her through the forest.

It was a beautiful fall day. The leaves were just beginning to turn a golden brown, the temperature was between 65° and 70°, and the sun was shining. Lucy crossed the dam of a pond, walking slowly through weeping love grass and entered the forest. As we followed her in the forest she would occasionally run away from us, then come back, then walk slowly forward to disappear and reappear behind us. We could not predict the path that she would take; we simply followed her as best we could. All went well for a while. Natasha was delighted to see a chimpanzee walk through a forest, and to be with Lucy in her natural setting is indeed a sight of great beauty. She is so completely at home in the forest, so completely relaxed, one feels a sense of awe at how well adapted she is. The forest holds no fears for her, only delight. Her movements are a flowing ballet of grace and dignity. She never loses her balance or seems uncentered, whether she is walking, climbing, sitting, or standing. Natasha was delighted.

We walked through about a half a mile of woods and were coming toward a small pond when Jane excused herself and said she would join us later. Jane walked off in the direction from which we had come, and Natasha and I continued to follow Lucy toward the pond. We circled around the edge of the pond when Lucy climbed a small hill, the top of which was covered with oak trees. Lucy delights in running up and down a dirt bank, particularly when there are trees at the top of the bank. She ran up and down half a dozen times, getting more and more excited. I noticed her excitement, but did not anticipate any problem as she was having such a good time. I even signaled her in the American Sign Language of the Deaf, "Come tickle Daddy," thinking she might leave the trees and the bank and come to me for a quick kiss and caress. She continued, however, to amuse herself as Natasha and I approached the bottom of the little hill. It then happened so quickly I did not know it until afterwards. Lucy bit Natasha, without rancour, anger, hostility, rage, or any of the kinds of interpersonal hassles which usually produce a bite. The tragedy was that Lucy bit her simply by being herself in the excitement of running down the hill.

Lucy ran down the hill toward Natasha. Natasha saw her coming and thought Lucy was going to jump into her arms. She was recovering from the flu and did not feel strong enough to catch a chimp running full speed into her arms. Automatically, she took a step towards me and put her arm around me. In the excitement of the moment, her aggression mobilized by running, Lucy misinterpreted the movement and bit her forearm. It was a small bite, one-and-a-half inches long and about one-fourth inch deep on the upper surface of the forearm from her canine teeth. Natasha also had several deep bruises on the underside of the forearm from the molars. But the bite was enough to be painful and it certainly was discouraging to me. I hated Lucy at that moment. Natasha is a sensitive woman, a friend and a guest, and I felt responsible for the incident since I knew it was risky to go to the ranch. And, of course, I loved Lucy too, so there was simply nothing I could do to ameliorate the situation in any way and I felt horrible—as did Jane when she returned a few minutes later. Yet it is not hard to understand and might even have been predicted had I been more aware.

Lucy, seeing Natasha move toward me, the dominant person in her life space at the time, when considerable aggression was already mobilized by running down the hill, simply bit her slightly. Whenever one chimp charges another, a mild blow or bite is almost always inflicted if the other runs.

Natasha herself behaved as the person she is—she was neither anxious, punishing, nor artificial. She did not scream, she did not run, she simply said "Look," and showed me her arm. She had recognized that Lucy was not mad at her, that had Lucy really been mad she would have had a much more severe bite. (Actually, had Lucy been attacking hostilely Natasha would have lost the muscles of her forearm.) Nor was she rejecting of Lucy. But I was so angry at Lucy I could hardly talk to her and I wanted to take Natasha to the hospital right then because I was afraid of an infection. Chimp bites sometimes contain exotic bacteria, or common bacteria in exotic combinations, and often require more than conventional broad-spectrum antibiotics. But Natasha felt that this was the chance of a lifetime, and she wanted to continue the walk through the woods to watch Lucy's, as she put it, "beautiful movements." I felt I had seen enough of Lucy's "beautiful movements," but I said nothing and we continued our walk with Lucy through the woods. Later Jane and I made the decision to keep Lucy and Natasha apart during the next few days, and never to take people into the woods with Lucy again.

There were two interesting occurrences after Natasha's visit which stand out in my mind.

During her last group Natasha had asked the group to do a self-image exercise. This consisted of each person drawing a life-size picture of himself on long rolls of white wrapping paper. The group members then did Gestalt work with their self portraits, associating to different areas of the drawings that seemed to have psychological significance or acting out various aspects of themselves in the drawing. After the group Jane and I took Natasha to the airport. After saying good-by to Natasha at the airport, we drove home feeling relaxed and at the same time emotionally drained since we had had a very intensive week-end experience. We had not cleaned up the house. There were crayons on the floor as well as large rolls of white paper. When we let Lucy down from her roof-top room she rushed into the living room and began to explore the chaos as she always did after other people had been there or after the furniture had been moved. She ran around the room smelling ash trays, empty glasses, left-over food, and other debris, and then noticed the large rolls of paper and the crayons. She quickly unrolled the paper, grabbed a box of crayons, dumped them out and started to draw. She drew very intently and was quite absorbed in her work. Her drawing was interesting because she seemed to be drawing circles which chimpanzees allegedly are unable to draw.

Since the workshop was a movement and Gestalt workshop, there was a lot of music. Natasha brought her own and played it on Steve's stereo set (which frequently went kaput in the middle of it). She had several records of Moroccan and African music which appealed to me, and after she left I bought some records of African music. They are not the same records that Natasha had, but they were recorded in Africa by various tribes dancing to their own music made with African instruments. The next night when Jane and I had just finished dinner I put on the music. Both of us spontaneously started dancing to the pulsating drum beats of the African rhythms. Lucy stood erect on her hind legs like a human being in the middle of the dance. She did not move to the music, but she turned around watching our wild gyrations and clearly seemed to think we were crazy. Many times I had let Lucy listen to my tape recorder as I dictated the first draft of this book. She always showed disinterest. When the music played, however, she became quite interested in the cassette recorder, trying to push the buttons to turn it on and off and to open the magazine.

I also had a cassette recording of African wildlife sounds. None of them was chimpanzee sounds, but there were sounds of a leopard, African lions, elephants, vervet monkeys, and a rhinoceros snorting. After Jane and I finished dancing we put the African animals on to see Lucy's reaction. She was fascinated. She put her ear close to the recorder, tried to take it away from me and tried to open it, and was obviously fascinated with the jungle sounds. I found this interesting, as it gave me elaborate fantasies about Jung's concept of racial memories. Lucy, of course, has never heard a leopard bark in a jungle or any other jungle sounds, except perhaps of jungle birds since my parrots make many sounds. She clearly was fascinated nonetheless.

Tickle-Chase, Blind Man's Bluff, Dress Up and Keep-Away are some of Lucy's favorite games. Lucy has enormous strength, but maintains distinction between play and fighting in roughhouse games.

CHAPTER FIVE

Games Lucy Plays

Y OUNG CHIMPANZEES are delightful playmates. They rarely seem to tire of play. Then, when they do tire, they need rest only briefly before starting again. Jane Goodall brought back from Tanzania many movies of chimpanzee youngsters at play, chasing one another, playing hide-and-go-seek, wrestling, biting one another playfully, and the like. While the adults do not play as much as do adolescents or very young chimps, they are nonetheless playful, too. And there is something about Lucy, and the absurdity of raising her as my daughter, that enhances my sense of humor. I can enjoy being ridiculous, or being as a child again with Lucy in ways I could not tolerate in myself before living with her. For example, it amuses me to call her "Precious Darling" when she could tear me limb from limb if she wished, or to whine "Oh, how could you do this to me?" Also, Lucy takes such a spontaneous joy and delight in the simple conveniences of civilization, as well as in nature, that it becomes a pleasure to give to her and play with her. Listening to the ticking of "Daddy's watch," tasting "Daddy's drink" or blowing out Jane's match when she has just lit a cigarette—all such simple actions are great delights to her, and her joy spreads throughout her family.

Lucy's joy and delight in uninhibited, spontaneous play are greater than that which most humans are able to exhibit. Sometimes she plays with her toys, more often with people, and sometimes with Steve's dog.

Nanuq

Nanuq is one of Lucy's favorite playmates. Nanuq is a female chow, a large handsome dog selectively bred for centuries for its strength and courage, qualities which made it a favorite dog of both Sigmund Freud and Konrad Lorenz. It takes strength and courage to play with Lucy now that she is grown, and Nanuq has both qualities and now has a beautiful relationship with Lucy. Like all meaningful relationships it did not spring full blown from the head of Zeus, and it certainly was not love at first sight. Their friendship developed gradually, with many ups and downs.

Nanuq (pronounced "Nanuk") is an Eskimo word meaning polar bear. Steve named her Nanuq when I gave her to him as a puppy. The name was meaningful to him because she looked like a small bear, though she had reddish fur, and Steve's favorite sport is mountain climbing above the snow line. Since early childhood he has been interested in the adventures of polar explorers, and as soon as he became a teen-ager he started making back-packing trips, then started to learn mountain climbing. Steve's idea of a great vacation is to spend a week or so on top of a mountain, preferably in an ice house he made himself. He feels such lonely hardships help him develop himself as an independent person and also to get away from the hassles of planning a career and going to school. After meditating on a mountain for a week he always comes back refreshed and ready to face the world again.

Lucy's first approach to Nanuq was aggressive, as it had been when she first saw her cat. Her hair stood up, she barked, her most savage sound, swayed from side to side, and looked ready to attack. At that point civilization intervened to sublimate and channel Lucy's spontaneous aggression. "Be a nice girl," I said, and Jane added "nice puppy" as she petted Nanuq. Jane took Lucy's hand and gently stroked Nanuq's back to show her it felt good to be "nice to the puppy." Very quickly Lucy's aggression disappeared, and within an hour or so they were friends, but Lucy was so strong and Nanuq so little at the time that we could not let them play together without supervision. When Lucy's play would get too rough we would have to stop her, and that would frustrate her, so that her reaction to Nanuq would suddenly turn hostile for a brief period, and Nanuq would run across the room in fear. At other times, in other moods, her approach would be friendly. Gradually they began to become more consistently friendly. When Nanuq was two years old, Jane and I took a trip to California together and Steve spent a week at the house with Lucy and Nanuq. By the time Lucy and Nanuq had spent a week together they were fast friends.

It now is a joy and delight to see them play together, at least it is when I can conquer my fear that they will get carried away and wreck the house.

Their play is very fast moving and very rough and it illustrates a wonderful connection between love, play, and aggression. At least for them (if not for human beings) love involves play and play involves aggression. Without aggression there is no play and with too much aggression play turns into hostility. They will chase one another about the room, Lucy jumping over couches and chairs with Nanuq rapidly behind her. As fast as she is, Nanuq can never catch Lucy. Then Lucy will turn around and Nanuq will crouch and growl, and they will try to stare one another down. At other times Nanuq turns tail and runs with Lucy right behind her, only to turn and take a stand from which Lucy then laughs and runs. It is clear to the observer that they are not trying to hurt each other, that it is a game; and it is a game with its own rules. After four or five minutes of chase and being chased they will tire and sit on the sofa together. Then Lucy will caress and stroke Nanuq, petting her gently on the head and running her hands along Nanuq's back.

When their relationship first started, Nanuq had difficulty reading Lucy's signs and interpreted her aggressive play as hostility and sometimes became frightened. Now Nanuq seems to understand what she wants when Lucy does such things as put her forearm in Nanuq's mouth. Nanuq will then gently bite down and Lucy will laugh and laugh and laugh, then take her hand away and run, and Nanuq will leap off the couch in rapid pursuit. When their aggressive play turns hostile, as it sometimes does, there is no question of who wins.

It is Lucy's superior intelligence and tool-using capacity that always turns the tide. Occasionally, for example, they will be chasing one another through the house. Steve and I will be sitting in the living room watching, and we have seen the sequence many times. Nanuq will chase Lucy. Then Lucy will turn and chase Nanuq. They will run around and around the chair, then over the chair, then over the couch in a seemingly endless chaos of running play. Then at some point, for some unknown reason, perhaps because she gets carried away with aggression, Lucy decides to pick up a chair and holds it over her head. Now that she has learned, Nanuq instantly stops the play. Before Nanuq learned to "read Lucy's mind," there were several instances where Lucy held a chair over her head and then threw it at Nanuq or, like a circus lion tamer, held the chair in fornt of her and slowly backed Nanuq into a corner and would have hurt her had we not stopped the play.

Steve and Nanuq have a heart-warming relationship. Steve expresses his love for Nanuq by taking wonderful care of her, by teaching her to be well-behaved and disciplined when appropriate, or to run with spontaneous joy when in open country. She will come when called, heel, or sit on command, is friendly to others, but would protect Steve with her life. She and Steve have had some touching moments together, lost in a blizzard in the mountains of New Mexico or living for a week in a snow house Steve made atop a Colorado mountain. Yet it is the love af a young man for his dog and, in this father's eyes, it is different from his love for Lucy. I do not mean there is more or less love in one relationship than another; it is just that it is different. Nanuq is Steve's dog; Lucy is not his chimpanzee—or mine or Jane's. The thought of owning Lucy, even as a beloved dog is owned, is to me like owning another human being, and that cannot be done, and could not be done even when the institution of slavery flourished. Minds and hearts have a way of remaining free, if they work and wish, even on the rack.

Steve and Lucy's love for one another is more like the love of equals, of one sentient being for another. There is no possession in it, often they do not even agree, and Lucy frequently "does her own thing" and disobeys Steve. I have seen them threaten and fight to the point that I have feared for Steve—though I never interfered so that they could work it out for themselves. The love between them is the kind of love Martin Buber characterized as a "face to face" relationship, generally a tender, though sometimes tough and always real confrontation between two conscious beings, each caring for the other while simultaneously aware of and respecting his own uniqueness.

Toys and Games

Lucy has her own chest of toys, a large plastic basket which stays in the living room. Going through it this morning (October 29, 1974), I found, in order, a small plastic suitcase; a woman's hairbrush; a plastic telephone which rings; a woman's leather purse; a mirror; a book of bird pictures; a plastic purse; a set of barrels which may be taken apart and inside each barrel there is a smaller barrel; a leather wineskin; a plastic banana; a plastic carnation; a woolen skier's head and face mask; a pair of tennis shoes; a Rand-McNally Junior Elf book with pictures of kittens; a pair of gloves; three small dolls; a pencil-type flashlight; a plastic orange; crayons and white paper; a lady's compact; a briar pipe; a plastic drinking glass; another hand mirror; a comb; another brush; a set of plastic bolts which may be bolted onto a form board; a deck of playing cards; a second small plastic pipe; a square foam pillow; a rubber egg; a Spiro Agnew watch; a small lock and key; the arm and leg of a dismembered doll; a heavy metal buckle which had been removed from a leather belt; a harmonica; the dismantled remains of her kiddy-cart, which she used to drive about the living room; a little music box which plays "Raindrops Keep Falling on My Head;" several tiny metal Volkswagens; a plastic spoon; and a motorcyclist's helmet. She has other toys, too, which we keep locked up—fingerpaints are particularly lethal when we are not supervising her because she uses them to write or paint on the walls.

Tickle-Chase

Her favorite game is tickle-chase. It consists of chasing Lucy, catching her, and tickling her. Or, simply tickling her. She will initiate this game in many ways: by approaching, turning around and bending over, looking at us through her legs and stretching one leg toward us, shaking the foot. We are supposed to grab her and/or tickle her. When tickled, she laughs, a soft, panting, gutteral laugh. When she's had all the tickling she can stand at the moment, she runs, looking back over her shoulder to incite us into chasing her. Although she could easily outrun us, she almost never runs so fast that we cannot catch her to tickle her. Or, since she learned the American Sign Language of the Deaf, she may incite the tickle-chase game by looking at us, smiling or laughing, while taking the forefinger of her left hand and running it rapidly across the back of her right hand to sign "tickle-chase." Being tickled is the most powerful reward one can give Lucy. She likes being tickled hard and we have never been able to do it so hard that her laughter stops with pain. I have used all my strength to dig my fingers into her muscles, and the harder I dig the more she laughs. Occasionally both Jane and I will chase her and when we catch her I will grab her hands or her feet and Jane the other extremities, lift her and swing her by her hands and her feet. She dearly loves this, though it seems to get so pleasurable for her that she can't stand any more of it, and she then contracts her knees and her elbows and pulls us into her and escapes. Even with two of us holding her, since she has been full grown we have never been able to hold her so that she could not escape.

Lucy can tickle herself, too. I have never seen a human being who could tickle himself, even if he were a ticklish person when being tickled by someone else. He still cannot tickle himself. But Lucy can send herself into peals of laughter by digging her fingers into her own muscles, usually those of the jaw, arms, or legs. It is a real difference between humans and chimps.

Wrestling

While tickle-chase sometimes involves wrestling for a moment or two when we catch her, I classify wrestling as a separate game because sometimes Lucy and I or Lucy and Steve wrestle in earnest. At these times, Lucy grasps the idea that the game is to hold the other person down or to keep him from escaping, and then she wrestles in earnest. It is utterly amazing how powerful she is and how quickly she moves. I have had her down with my one hundred and seventy-five pounds on top of her eighty-five pounds, my shins covering her thighs and my hands on her forearms holding her on her back. All of a sudden she stops laughing for a second and I find myself off her, either thrown backwards or simply lifted up and off. Since she has in effect four hands it is impossible to really hold her down when she is wrestling, unless she wants to tease us into thinking we've got her. When he was in high school Steve was a state champion in Judo. Nonetheless, he cannot win when he and Lucy wrestle. When she gets into it Lucy can move so fast that it is impossible sometimes to see her movements. Now that she is an adult, her wrestling with Steve sometimes gets so intense and rough it scares them both, and Lucy will then shriek and walk away, which I interpret as a fear that she will bite Steve. Steve also is afraid of being bitten now, and he is careful to wrestle with Lucy only when she is not in estrus.

Blind Man's Bluff

When Lucy was growing up she took a great delight in covering her face and attempting to walk blindfolded. Sometimes she would put her ski mask on backwards so she could not see, then start walking slowly across the room bumping into the furniture and laughing when she did so. Sometimes she will wear her motorcyclists helmet over her face and turn rapidly around, making herself dizzy, and then try to walk through the living room, collapsing with laughter each time she bumps into a piece of furniture.

Playing With Water

Jane Goodall observes that chimpanzees in Africa do not like to play with water or get wet. They jump across rather than wade through shallow streams and they seem acutely unhappy, even miserable, when it rains. They do not bathe, but stay clean by grooming themselves and one another, and by a natural process of regeneration of the cells of the skin. Lucy is not this way. Though we do not bathe her, and while she does not swim, she does play in water. She will stand in it and slop it all over her; kick and slap the water. She takes a particular delight in playing with a garden hose. We have seen her plan the whole sequence in advance; that is, run out to a hose which is not connected, which is coiled up on the lawn. She selects the right end without any trial and error, carries it to the faucet and screws it onto the faucet, turns the handle, then runs to the other end of the hose to play with the stream of water. Sometimes she puts the stream in her mouth, putting one finger over the end so that she receives a squirt under pressure. At other times she caresses her genitals with a stream of water, masturbating as human female children often do. Sometimes she puts the stream of water into her vagina; other times she will simply squirt it onto her clitoris.

To the left of our front porch there is a large oak tree which Lucy loves to climb. Many times we have seen her take the garden hose up into the tree and sit or stand on a limb and watch the stream of water splashing down. Similarly, there is a hose in the courtyard which has a spray nozzle at the end used for watering the plants. The water comes out under considerable pressure. The nozzle is the kind with a handle which can be squeezed to make a fine spray or a small direct stream of considerable force. Lucy loves to play with it. We do not allow her to do this unless we are there and even then we discourage it as much as we can because she frequently sprays the parrots. This seems to be on a devilish sudden impulse. She has the hose in her hand spraying her genitalia, suddenly she notices the parrots, and a second later has sprayed them.

Incidentally, her attempts to devil parrots may be the cause of an accident we had on October 28, 1974, which changed our whole life. We discovered our house was no longer "chimp proof." Lucy knocked out the glass door between the living room and the courtyard. It happened during the night and we did not know it until the next morning. The glass was shattered, there were glass shards all around, along with considerable blood. She had cut her hand and arm in three or four places. In attemtping to understand it, we examined the courtyard carefully, but the only thing unusual was that one parrot was loose. It was the parrot that Lucy does not ordinarily like (perhaps I should say the one she likes least, because she really does not like any parrot), and it could be that she was hitting the glass window to intimidate the parrot who should have been in his cage. In any event, it was the end of an era for it meant that Lucy could no longer be left alone in the living room, or for that matter in any room in the house except her own room as there are glass windows in all other rooms.

Running Away

Running away is a great game for Lucy though it scares us half to death. Lucy is very careful to watch us as we lock doors behind us, and she is gone in a flash should we make a mistake and fail to lock a door behind us, leave a window open, or leave our keys lying about. Although I carry fourteen keys on a ring, she has on two occasions escaped by picking up the key ring while I was taking a bath or reading and letting herself out. She easily can select the correct key from the fourteen on my ring and open any door, regardless of the type of lock, if she has the right key. If one key does not work, she tries for a few moments, then gives up and selects another key. On two other occasions she escaped because of failures of communication between Jane and me. For example, one of us would be bringing Lucy down from her upstairs room while the other one would have unlocked the door to go outside and left it open, thinking Lucy was locked in her room. When she ran away, we were always panicked. Although we have five acres, a much-traveled highway runs along the north boundary and there is a housing development on our south boundary. We also have neighbors living on one and two-acre lots on both the east and the west. Thus the population density is much too high to support a full-grown chimpanzee running wild for very long, and we have been terrified that she would be hit by a car or that neighbors would panic and shoot her. On each of the occasions when she ran, we immediately phoned every friend or anyone who was connected with Lucy in any way and scoured the woods. Occasionally Lucy would seem to double back on us or to "track" the trackers through the woods. Once, after two hours she was found sitting in a clearing about half a mile away. These times frightened me, but Lucy seemed to be quite content and to enjoy them as though it were a game.

Dress Up

Lucy loves to dress up. She will adorn herself with a blanket or any interesting article of our clothing, wrap it around herself, take it to a mirror and watch herself in the mirror as she preens and postures in Jane's scarf or blouse, or my hat or shirt. She particularly likes to put on Jane's or my clothes, though occasionally she will voluntarily put on her own. We have tested this many times. For example, a favorite item is a pair of my boots. If I sit passively and let her, she will unlace them, take them off of me, and put them on herself. She liked them so much I bought her a pair of boots of her own, and she completely ignored them. But she will always wear mine, clomping around the living room floor in them with great delight. She has shown no preference for "girls" clothes over "boys," but dresses up in the clothes of either sex with equal delight. And she loves to unbutton Jane's blouses or my shirts, frequently as a substitute for the grooming behavior wild chimps show one another.

Keep Away

Lucy developed this delightful game herself. It consists of grabbing a precious object and running away with it, but not completely away and out of sight. Rather, she stops just out of reach and laughs at our attempts to recover whatever prize she has stolen. And Lucy has an uncanny knack for knowing just what prize is most highly treasured and will get her the greatest emotional response. If I am reading, for example, she will grab my glasses and run laughing through the house as I chase her. Or keys—she knows that with keys she can run away and that we are helpless without our keys. Even going to the pantry or the bathroom requires a key in Lucy's house. So she is especially fond of stealing my key ring and playing keep away with it. It is impossible to catch her and she knows it. Even if she stops in the living room she always wins at this game. Not only can she run faster than we can but she can leap over sofas or chairs without slowing down at all, while the furniture is an obstacle to me. The only way I can get back my keys, glasses, or whatever is to become *genuinely* exasperated, to have actually reached my emotional limit. When I have become so frustrated that I am ready to tear my hair out, then Lucy recognizes my change in mood and will stop the game. She then may hand me the prize, though more frequently she drops it wherever she is at the moment and watches me intently as I recover it. But until I reach my emotional limit, no amount of "Please stop," or "Okay, Lucy, give me my keys," or "Be a good girl, give Daddy his keys" will have any effect. She will simply tease me with them, bring them close to me, hand outstretched, but turning and running away when I reach for them. When she sees that I am really angry Lucy stops. Often she then will request reassurance by whimpering, or coming close and touching hesitantly.

For awhile Lucy had a friend named Sue Savage. Sue was working on a Ph.D. in psychology and was one of the people teaching Lucy the signs of the American Sign Language of the Deaf. She was a young, attractive girl in her late twenties. When she visited Lucy, she often brought a football helmet because Lucy dearly loved to steal and play keep away with it. Sue wrote Jane a note describing Lucy playing keep away which is worth recording here.

"We went outside and Lucy took the helmet, generally carrying it high on her arm, above the elbow. All she wanted was to play keep away. Whenever I would sit too long for a rest she would come and tug at my hands. If I still sat she would sign "tickle Lucy." If that did no work she would start thinking up something to sign. "Hey, you, you, you, you, tickle Lucy:" or "Boy, you please tickle Lucy:" and "You come, you come, come tickle Lucy." And for emphasis she would grab my hands and make the tickle sign.

Outside the keep away game is much more vigorous and intense—but strangely she is not so rough. It might look as though she's going to yank me down and bite my arm off but the actual grab or play bite is very gentle and then she's off. I feel that this has something to do with the added amount of space in which to maneuver and play and may be a key to why play sometimes turns to aggression in colony chimps. The keep away game is fascinating because, for one thing, it can be turned off so rapidly—almost as though we were playing it by a formalized set of rules—like football is played. After all, most of our favorite sports are nothing more than highly ritualized forms of keep away. She will defend that football helmet with all her might when I try and take it from her; but with a slight change of expression and tone of voice all I have to do is say "Lucy' let me see that for a second," and she promptly hands it to me. One of the rules seems to be that the height of aggravation to the opposing opponent is to allow the object to roll away from your grasp, but to recover it just before it gets in his hands—just like the hearts of football viewers skip a beat when a fumble occurs."

Kendo

One day Lucy and I invented a game which for lack of a better name I'll call Kendo. Kendo is a Japanese martial art in which the opponents fight with staffs, usually long poles of hardwood. That morning I had been working in the bird flights repairing perches and I had two oak saplings about six feet long which I was going to use as perches. All the branches had been removed so that they were straight oaken staffs about an inch in diameter. These were on the front porch where I was standing, watching Lucy play. She was running around in circles. It was a lovely fall day and I enjoyed watching her enjoy herself. Suddenly I remembered that the male mating display involves the use of sticks or branches. The male will often pick up a stick or tear off a branch and hold it over his head shaking it or waving it at the female in estrus. Lucy was in full bloom that morning and had she been in nature she would have instantly copulated with any male who made such a gesture or otherwise indicated his desire to mate with her. To see what would happen I grabbed one of the staffs and waved it over my head, menacingly, I thought, as I approached Lucy with it. She showed not the slightest tendency to mate with me. She instantly picked up the other stick and approached me doing the same thing, waving the stick over her head at me. I then hit her stick with mine in the manner of sword fighters, or perhaps pirates brandishing cutlasses. Lucy quickly grasped the idea and we had invented the game of fighting with our sticks. Occasionally Lucy would grab her stick with both hands and parry my blow by holding her staff parallel to the ground so that I would hit her stick between her hands. She would then hit me, often very hard, with her stick while charging. This game never lasted very long though it was fun while it lasted. But it was very short because it was so vigorous and so aggressive that I would tire within two or three minutes, often less. Or Lucy would end it by getting so carried away with aggression that she would forget the game aspects and drop her stick and charge and grab me with her hands and put her mouth over my leg or hand. The game had the effect of defining the oak staffs as play objects to be prized because many times after that, as we approached the front porch, Lucy would see them and grab both of them and run, playing "Keep away." Or she would grab one, brandish it at me, and then sign "Tickle-chase Lucy."

In other words, Lucy has an uncanny knack for knowing how to start an aggressive game when she wants to play one, even if I do not. For example, after a day at the office I usually like to relax with a drink and read the newspaper. If Lucy wants to play and I say "no" and she really wants to play bad enough, she will steal my glasses, grab them and run; or before I start the drink, steal it and run, forcing me to chase her.

CHAPTER SIX

Creative Masturbation

For a long time anthropologists thought that man was the only living organism to use tools. Then Jane Goodall, carefully integrating herself into a group of wild chimpanzees in Tanzania, saw them not only use tools, but construct tools as well. For example, she observed chimpanzees chewing leaves and then using the semi-chewed leaves as sponges to pick up water. And she also saw them using a stick to obtain termites from a mound. The chimpanzees would often strip the twig or stick of leaves, then stick it into a mound, remove it, and lick off the termites. These are extremely simple but clear examples of tool use. Actually, several other animals use tools. The California sea otter will use a rock as an anvil, carrying it on his chest, pounding an abalone on it until the shell opens. I once had an African gray parrot who used water as a tool. He would soften a monkey biscuit in water before he ate it.

Lucy's Use of Tools

Lucy became proficient and creative in the use of many different tools. It is Lucy's use of tools that most dramatically illustrates her potential for becoming human through social learning. Her resemblance to man, particularly to "natural" man at his most uninhibited, is striking.

We have seen Lucy use correctly the following tools: spoon, knife, fork, cup, dish, shoes, cigarette lighter, matches, water hose, bucket, shovel, rake, vacuum cleaner, mirror, book, crayons, pencil, ladder, pliers, wrench, screw driver, wastebasket, ash tray, electric light switch, tricycle, comb, brush, key—to name but a few.

She learned how to use a screwdriver by seeing me use it once. Since then screwdrivers have become a highly prized bit of contraband and we have had to keep them hidden under lock and key as Lucy can dismantle the electric fixtures in the house. The first time she did this she received a shock. She learned on one exposure, and the next time she took the plug apart, she did so without getting shocked. Once she removed the kitchen doors by unscrewing the hinges.

She has learned to use pliers very effectively. We have seen her carefully grip her labia to pull them apart with the pliers, and she did so without hurting herself by squeezing too hard. Outside in the summer she loves to play with any different tools, and often carries them into trees with her. We have seen her carry into the tree a water hose, buckets, footballs, basketball, mirrors, etc.—anything which she can sit comfortably on a limb and play with.

We first discovered Lucy's capacity to use tools when she was only three, and it was one of the most dramatic experiences of all.

When we built the house for Lucy we put locks on all internal doors. In other words, bathroom doors, pantry doors, doors between living room and bedroom, etc. all were keyed and locked as though they were doors to the outside of the house. We had them all keyed the same way, too, so that one key would open all internal doors and another key would open all doors to the outside. Also, the locks were designed so that they could be locked from both sides so we would have greater control. In other words, if Lucy were in the living room, we could lock the door keeping her out of the bedroom. Or if she were in the bedroom we could lock the door so that she would have to stay there and could not get back into the living room. With each lock set that we bought there came two keys. So for a while after the construction was completed there were many keys laying around the house. They tended to get misplaced very easily as we did not need them all, and at the time we did not realize the danger.

When the house was finished we locked Lucy into her room and left for work contentedly. We were happy in the knowledge that she had almost a thousand square feet to play in, that her room was filled with toys, and that she would be safe and happy. When we came home the house was in a mess. The refrigerator had been opened and one bite taken of practically everything in it. Frozen food was scattered around the kitchen defrosting. Liquor was spilled, bottles were broken, and needless to add, Lucy was loose in the house. Jane and I immediately blamed one another. I said that she had not locked her in securely, and she said it was all my fault. The next day we took added precautions. We each checked to see that Lucy had been locked in, and then we left for work. When we returned that evening we found that the same thing had happened again; there was chaos everywhere. We then knew, of course, that there was some kind of leak in our security system, but we could not figure it out. We both felt rather stupid. Lucy appeared to be out-thinking us. The next day we each locked her in and drove off as if to work. However, we circled around the section of land in which we lived, and came back surreptitiously about ten minutes later. Jane sneaked in one door and I entered by another. I caught Lucy redhanded opening the door to her room with a key. The minute she saw me she looked guilty, and plopped the key in her mouth. We realized that she had appropriated a key, hidden it in her mouth, having understood its use perfectly and had unlocked herself.

I felt not only stupid but hurt, "after all I had done for her," that Lucy would want to run away from her new and lovely room. Since that episode we have been fully aware of Lucy's capacity to use tools.

Her use of tools can sometimes be frustrating in other ways. Lucy can put a nut on a bolt with a wrench so tightly that she sometimes breaks the bolt. Or conversely, it is frustrating to me when working around the house to put a nut on a bolt with a tool only to see Lucy unscrew it with her fingers.

Unlike chimpanzees in nature, Lucy's drinking habits have been modified by her tool-using capacity. In nature, chimps lower their face to the water, extend the lower lip, and lap up water. Lucy drinks from a glass. When thirsty, she will go to a cabinet, open the cabinet which is latched by pressing on the cabinet door and suddenly releasing it, pick up a drinking glass, turn on the faucet, and fill it with water.

One time Lucy locked me out on the front porch. In the summer I often enjoy doing yoga early in the morning nude on the front porch. One day Lucy got up early, left the bedroom for the living room, and went back to sleep on the couch in the living room. Without thinking I opened the door to the front porch, stepped out, left the door open, and started doing yoga. I was standing on my head when I noticed Lucy standing at the front door. Her hand was on the key which I had left in the lock. She quickly closed the door, turned the key in the lock, and I was locked out of the house, nude. Lucy then went to the window, a large picture window which separates the living room from the front porch, and watched me as I furiously asked her to open the door. She ignored my commands so I pleaded, but Lucy continued to ignore me and I had to break into the house through a window.

Creative Masturbation

I have always felt that Lucy's most ingenious use of tools was to use them to explore herself and to masturbate.

When Lucy was almost three years old I saw her exploring her genitalia for the first time using tools—a mirror and a pair of pliers. It was a small pocket mirror and it was on the floor. Lucy was squatting over it, watching her genitals in the mirror as she held the labia apart with the handle of the pair of pliers. I was absolutely fascinated and called "Jane, come here quick!" Jane was in the bedroom and walked into the dressing room. Lucy continued to explore herself as I told Jane, "My God, look what Lucy's doing." "Of course," Jane said, "that's just the way little girls do." I suppose it is, but it was my first contact with the idea that female chimpanzees, like human females, would have more use for tools in exploring their genitals than would men, since female genitals are more hidden inside the body. Since that time Lucy has continued to explore herself and to masturbate with several different tools.

The next one was a pencil. Before Lucy reached sexual maturity her masturbation was exclusively clitoral and she often used tools to stimulate the clitoris. One morning as I sat on the couch beside Lucy I noticed that she had a pencil in her hand and was stimulating her clitoris. She would gently poke around the clitoris, and then slowly rub it with the pencil. I watched her for a few seconds as she continued to pleasure herself in this way. She then noticed that I was watching and she handed me the pencil. I was dumbfounded for I thought that she was inviting me to masturbate her. However, after handing me the pencil she got up off the couch and went to her toy basket and began to play with something else.

After Lucy began to ovulate, her masturbation changed, for it became both clitoral and vaginal. According to most surveys, the human female generally prefers clitoral stimulation during masturbation. However, after she started to menstruate Lucy masturbated both with clitoral and vaginal stimulation—usually stimulating both vagina and clitoris at the same time and with equal frequency. She masturbates frequently (at least once a day) whether or not she is in estrus. One of her favorite tools with which to masturbate is the vacuum cleaner. She will play with it for hours, turning it on and off, putting the suction to various parts of her body, including her genitals. She is also fascinated with small objects disappearing into the tube. While Lucy seems to understand perfectly well the cleaning function of the vacuum cleaner, it is clearly a toy for her. She runs it all over her body, chuckling with great delight, and puts it into her mouth and then onto her genitals, then reverses the sequence. The particular vacuum cleaner we have can be changed by a switch from suction to blowing. Lucy will sometimes blow air into her mouth or onto her genitals, and then reverse the machine, going from blowing to suction sensations at both orifices. One afternoon around five o'clock Jane and I were sitting in the living room when we observed this sequence of behavior.

Lucy left the living room and went to the kitchen, opened a cabinet and took from it a glass, opened a different cabinet and brought out a bottle of gin. She poured two or three fingers of straight gin into the glass. She then came back to the living room couch bringing her drink with her. From a coffee table she picked up a copy of the *National Geographic* magazine, lay down on the couch and sipped her gin while leafing through the magazine. About three to five minutes later she stopped suddenly as though an idea had hit her. She sat up straight and paused, put the drink and the *National Geographic* on the floor, jumped up and went to the utility closet at the far end of the hall, exactly fifty-seven feet from where she'd been sitting. She opened the door, took out the vacuum cleaner, brought the vacuum cleaner back to the living room and plugged it in a wall socket. She then removed the end brush from the long aluminum tube to which it was connected, turned on the machine, and applied the pipe to her genitals.

Lucy has plugged in the vacuum cleaner and turned it on. The second picture could be called oral foreplay. Below, Lucy is beginning genital exploration with the vacuum hose. Notice Lucy's intense involvement in the picture, upper right, as she applies the hose to her genitalia, swollen in estrus. Bottom right, having achieved orgasm, Lucy is enjoying a reflective moment before returning to her magazine.

She continued to masturbate with the suction from the machine until she had what I inferred to be an orgasm (she laughed, looked happy, and stopped suddenly). She then turned off the machine, picked up her unfinished glass of gin and her magazine, lay back on the couch and continued to drink while again contemplating the pictures in the *National Geographic.*

As I see such sequences of behavior, they clearly show a capacity for planning and foresight which no other animal but man exhibits. I also wonder if this is not the way we humans would be, had we not become so inhibited so early? The capacity to become inhibited is a double-edged sword, of course. Without inhibitions there are no controls, and without controls over our behavior we become creatures of blind impulse. For so many of us, however, the process of acquiring controls over behavior is overdone and we live restricted lives.

As clever as the use of the vacuum cleaner in masturbation is, I think it is not so creative as another form of masturbation we watched one morning, for it represented the maximum personal involvement.

Lucy lay on her back on the couch in the living room sipping coffee, which she held to her lips with one hand; the forefinger of her other hand was rhythmically thrusting deep into the vagina. With one foot she held a pocket mirror into which she stared intently, watching her finger in her genitals; the other foot held a copy of the *National Geographic* magazine open to pictures, the magazine propped against the back of the sofa. Lucy frequently moved the mirror from her genitals to her mouth, keeping a careful watch on both orifices while she sipped her coffee, looked at the pictures in the magazine, and stimulated herself! In a human being such activity might indicate a fragmentation of the personality, an inability to completely "be there" in one's own activity. This was definitely not the case with Lucy. As she thumbed through the magazine and studied the pictures she neither spilled a drop of coffee nor missed a beat as she thrust into herself ever faster and faster until she exploded into peals of laughter which I assume signified a "deep vaginal orgasm."

I have watched her masturbate in this fashion many times, and while she does seem to have her most powerful orgasms from vaginal penetration when her genitals are maximally enlarged, she is careful that the palm of her hand touches the clitoris as she thrusts her forefinger into the vagina. Her masturbation clearly supports the position of Mary Jane Sherfey (*The Nature and Evolution of Female Sexuality*) and not of Freud to the effect that "clitoral orgasm" and "vaginal orgasm" is a false dichotomy. Rather, the vaginal orgasm that Lucy has seems to be produced by the clitoral stimulation which results both from distension of the penetrated vagina and touching the clitoris with the palm of her hand.

To see what would happen on several occasions I attempted masturbation in front of Lucy. She ignored the entire process. One night I asked Jane to. Lucy also ignored her. However, when Jane attempted to masturbate me or when I touched Jane's genitals, in both cases Lucy immediately became very excited, grabbed our hands, and attempted to stop the genital contact.

Dangerous Tools

Lucy's capacity to use tools can be dangerous. One of her favorite tools is a cigarette lighter. Jane smokes, I do not, and Lucy likes to play with her lighter. Lucy lights it then blows out the flame. Since this can be dangerous, Jane started buying the kind which has a lever which must be held down before the flame will continue. This puzzles and frustrates Lucy, as she makes them click but the flame never comes. Then Jane will use it and allow Lucy to blow it out. It took a long time to teach Lucy to blow as chimpanzees do not blow naturally, but now whenever anyone lights a cigarette in her presence, either with matches or with a lighter, she promptly rushes over, sits down beside them, puts her arm around them, and blows out the flame. She almost never burns herself though she has done so, and while she is cautious about the flame she does not seem to be afraid of it. She has a similar lack of fear about other household fire. When she wants some tea or is in a hurry for her dinner she will climb up on the kitchen cabinet and turn on the burners, one after the other, watching them ignite from the automatic pilot light. If the teakettle doesn't seem to her to be full enough, she then takes it to the sink and fills it. Turning on the water and filling the teakettle is easily accomplished. The one thing that she does not seem to understand is the necessity for turning off the water after she has finished with the teakettle. On several occasions when she has wanted tea she has successfully accomplished the entire sequence, filling the teakettle from the sink, putting it on top of the stove, and turning on the burners.

Lucy's great strength and curiosity combine to make her destructive. This morning, for example, as I was reading a chapter I had written last night I did not realize that Lucy was playing with my tape recorder. I heard a few strange clicks but they did not register enough for me to do something. I could only see her back. When I finished editing the chapter I walked over to Lucy, put my arm around her, and then felt sick. She had dismantled the microphone from my tape recorder and this was Thanksgiving morning. I could not get it fixed and I wanted to dictate. I was angry at her, but felt also she was simply manifesting her intelligence and curiosity. She had not torn it apart but had just unscrewed everything possible. I slowly put the microphone back together hoping it would still work, plugged it in and dictated a paragraph to see if the machine worked and if Lucy reacted to hearing my voice. The machine worked but she did not seem surprised to hear my voice coming from the recorder, I presume because she had seen me dictate and play it back on many occasions.

One Sunday morning while Jane was sleeping late, Lucy and I went into the atrium. I sat down to watch the two baby Moluccan cockatoos. They were the first cockatoos I had ever successfully bred and were quite spectacularly beautiful, with their salmon-colored crests, blue eye rings, and white bodies which can turn to pink as they change the posture of their feathers—exquisitely beautiful birds. As I watched them, Lucy picked up the garden hose which had a spray nozzle. She adjusted the nozzle from a spray to a direct strong stream. With her left hand she picked up a brush which had wire bristles and a wooden handle. Holding the brush in her left hand with her right hand she squirted a stream of water at the feces on the floor of the cockatoo cage. She then scrubbed the floor as she had seen me do many times. She became obsessively involved and for periods as long as two minutes she scrubbed the same spot with her brush while squirting it with water. She reminded me of an obsessional housewife, compulsively cleaning the same spot repeatedly while ignoring the dirty spots six inches away. From time to time she would look at the birds and try to squirt them with water. I would have to yell "No, stop!" She would then return to her cleaning, but every few minutes she would try to squirt the birds again. After I stopped her twice she tried to spray the birds again, but this time she made it appear accidental. She looked away from the birds, then held the spray behind her back and sprayed them. She had something of an air of injured innocence about her when I yelled at her, since it had so obviously been an accident. Then she realized her efforts were hopeless, that I was going to protect the birds, and she started to masturbate with the stream of water. She held the nozzle in her right hand and squirted a strong stream of water to the floor and then slowly sat down on the stream of water until the full force of the jet was against her clitoris. From time to time she would reach back and touch her anus with the other hand, and then move her forefinger across her bottom into the stream which she alternated from spray to jet to change the sensation. She was not in estrus at the time and did not seem to masturbate to orgasm—though how does one know? The problem of the male observer external to the female chimpanzee trying to decide if she had an orgasm is as impossible as the anxious human lover trying to discern if he has "satisfied" his partner.

CHAPTER SEVEN

Can Animals Talk?

W HEN I was a little boy I knew with the utter certainty of the insecure mind that animals could talk. After all, my dog understood me and could talk back to me with her tail. She could tell me she was glad to see me when it was clear that the humans around couldn't talk about important matters, like whether they liked me or not. I could tell my dog about it and she would respond with a licking and tail-wagging if wordless affection.

A chimpanzee, of course, is a much more highly evolved organism than a dog, much more capable of human-like cerebration and communication.

Jane Goodall has recorded how chimpanzees communicate with one another by facial expressions, sounds, and gestures. For example, she describes a characteristic facial expression called a "play face" which is exhibited during play and accompanied by a series of grunting sounds very similar to human laughter. She also describes grins which signify fright or excitement; screams of terror, which may be gradually reduced to pouts or whimpers; and a barking which occurs when a chimpanzee is threatening.

Lucy makes many of the sounds and facial expressions described by Goodall. Since she does so without having

113

had any opportunity to learn them from another chimpanzee it strongly suggests they are species-specific, behavioral events somehow built into the biology of the chimpanzee.

Lucy uses them appropriately, too; that is, she utters them in the human equivalent of the social situation that elicits the same sound from wild chimpanzees. For example, when she is running to the refrigerator to steal a carton of fruit yogurt (a favorite treat) she food-grunts. Jane and I started to call her pant-hoots "good food sounds" before we read Dr. Goodall's book and learned they occurred in wild chimpanzees. Also, Lucy will greet me with a series of grunts when she is sleeping on the sofa in the living room and I awaken her as I return late from my office. It is interesting that this sound of greeting a loved one often is similar to the regular grunts that we call "good food sounds." Human love often uses oral metaphors.

When frightened Lucy displays one of the grin expressions Dr. Goodall described and squeaks, screams, or whimpers. Or if she is desiring reassurance from Jane or me, she will pout or whimper till we groom or stroke her comfortingly. To take one further example, when threatening another animal, a dog, cow or horse, she barks. On the few occasions when she has threatened Steve, or me, she has vocalized the bark—a sound she has never directed toward Jane.

Lucy occasionally makes sounds not described in Dr. Goodall's writings. These were always on occasions when something very unusual occurred. Once, for example, the floor drains in Lucy's room became clogged, and water and sewage began to gurgle and bubble, at first inside the drain pipe, then later backing up and flooding her floor. Lucy at first barked at the strange sounds coming forth from the floor drains, then as the drains overflowed she made totally new sounds. They may have been combinations of Waa barks and screams—I do not know for sure, as I only heard her do this one time. However, another observer, Sue Savage, did tell us that she heard Lucy make a new sound when the drains were gurgling preparatory to backing up, though we cannot be sure it was the same sound that I heard.

Jane, Steve, and I can make some of these sounds and have on many occasions. The sounds clearly have a standard meaning for her, as she responds about the same way each time she makes them or we do.

From the time she was two years old it was clear to us—though we could not prove it at the time—that Lucy understood many of the words Jane or I uttered. Though Lucy could not make our sounds, she understood many of our words and phrases, particularly if they were uttered in context. For example, Jane might fix four bowls of ice cream covered with fruit for dessert. Lucy might then ignore hers and try to get a spoonful of Jane's or mine—even though they were all alike. If one of us said "Eat your own," she then would do so. Or, if she clearly was thirsty, I could say "Get a cup and I'll fix you some tea," and she would open the right cabinet door, get a cup, and often bring me a tea bag from the pantry. In the winter we could say, "Let's go to the ranch" and Lucy often would become excited and grab her leash and jacket and run sit by the door, long before Jane or I had made a move to get ready. Indeed, Lucy so loves to play that it sometimes seemed she could understand any sentence dealing with games. Example: "Get the ball!" "Let's play chase." "Come tickle me." Or "Let's go climb trees." Any of these sentences always produced an appropriate response. Yet it remained for two comparative psychologists, Allen and Beatrice Gardner of the University of Nevada, to prove beyond doubt that chimpanzees could be taught a language and then communicate across species.

The Gardners began by contemplating the failure of Keith and Catherine Hayes to teach Viki to talk. For six years the Hayes had tried to teach a chimp named Viki to talk by shaping her mouth with their hands and encouraging her to blow air and imitate their sounds. Viki only learned four words and these she could not really articulate well—cup, mama, papa and up. The Gardners concluded that either chimpanzees were not intelligent enough to talk (which seemed unlikely) or that verbal language was an inappropriate medium, particularly since chimps have trouble blowing air and natural chimp sounds are so different from human sounds. Instead, they decided a gestural language would be more appropriate to the chimpanzee, and to try and teach a chimp communicative hand signals called *signs,* instead of sounds, they took the signs from the American Sign Language for the Deaf. This choice of gestural language was very appropriate because wild chimpanzees often communicate with hand signals, perhaps even more than with vocalizations.

The American Sign Language for the Deaf (ASL) is a complete language, often used for communication between deaf people, in which gestures made with the hands are substituted for words or phrases. Many of the gestural signs are iconics; in other words, the sign constitutes a visual representation of its meaning. For example, the sign for "drink" is made by touching the mouth with the thumb extended from the fisted hand. The Gardners learned the language so well they could give lectures in ASL.

The Gardners began with a one-year-old, wild-caught chimpanzee named Washoe. Washoe lived in a trailer and the Gardners and their students used only ASL to communicate in her presence, even when talking to one another. It quickly became clear that Washoe was learning the signs and using them correctly to construct simple sentences. For example, though she was taught signs separately, she combined two or more signs into such sentences as "Gimme tickle," her first sentence, and "Open food drink" to request that the refrigerator be opened.

The Gardners and Washoe became famous overnight because they together had demonstrated to science that cross-specie communication was possible. Given a language appropriate to their biological nature, gestural rather than verbal (because chimps find it difficult to make many sounds and use gestures to send messages in nature), chimpanzees can receive and return messages from humans. They could even construct sentences of their own and send them. It was a truly unique and wonderful discovery. In addition, it reaffirmed the faith of those who had felt that by studying the animal closest to man we would learn much about our own species.

Lucy and Roger

The Gardners had a student, Roger Fouts, who received his Ph.D. for his work with them. His dissertation was based upon a comparison of the different methods used to teach signs to Washoe. He knew ASL fluently, which meant he knew it well enough to give a psychology lecture to college freshmen in ASL. About five years ago he started working with Lucy. At first he worked with Lucy about an hour a day, and then twice a week for an hour each time; later he brought graduate students he had trained to teach her on intermediate days. Lucy was about five years old when he started out and, of course, she already understood a lot of the English language.

We cooperated with Roger, not only because his research had such obvious importance, but because we felt the broadened social contact would be good for Lucy. However, I did not feel that if Lucy acquired ASL it would enhance communication between us. We had lived so closely together for five years that all four of us could read one another's moods and feelings with ease, and most of the time Lucy understood and obeyed my spoken words. Nonetheless we opened our house and hearts and gave Roger our daughter to work with. I have always been glad we did.

Photograph by Nina Leen, *Life Magazine.*

Dr. Roger Fouts has taught Lucy over 100 words in the American Sign Language from which she makes her own sentences. Notice the necklace which is for her leash for occasional outings.

Roger was and is doing a lot of research, publishing like crazy, and his publications soon made Lucy famous. Communication across species in a language both humans and chimpanzees could understand caught fire in the public imagination. Lucy was visited by teams of reporters and television cameramen from German State Television, Canadian Television, the Columbia Broadcasting Company, and British Broadcasting Company, and the French Television Company. There were pictures of Lucy signing to Roger, or playing with her cat, or doing something human, in *Life, Coronet, Science Digest, Parade,* the *New York Times, Psychology Today,* the *Los Angeles Times,* and many local newspapers and television programs.

Roger was always very nice to me though I irritated him in two ways. I would be seeing people for psychotherapy in my office while he was working with Lucy in the living room, and a patient or I would have to walk through the living room to use the bathroom. Often this would interrupt the language lesson because Lucy would have to stop and see what was going on. Though Roger never said so I know I annoyed him at first with my questions. I would ask him: "Was she learning rapidly? How many signs does she know? Is she a good student? Does she know as many signs as Washoe? Does she learn a sign as rapidly as Washoe? Is she as smart as Washoe?"

Then I thought, what the hell am I trying to do, make a nice Jewish girl out of Lucy? The utter absurdity of life struck me like a blow as I realized I was projecting onto a chimpanzee competitive attitudes I had learned early in life (Be a success! Get ahead! Opportunity only knocks once! etc.). Here I was, pushing Lucy into competitive intellectual achievement as I had been pushed forty years earlier. So I laughed, perhaps with some bitterness, and recognized again Freud's dictum: "There is no time in the unconscious." I realized again how trapped I was, how shaped and molded by my childhood experience and somehow, as I became more aware of how caught I was I became, paradoxically, more free. So I stopped pushing

Lucy and trying to manipulate Roger into telling me Lucy was the smartest chimp in the world. I started liking Roger even more and appreciating his work—which I had not at first because I felt: "Of course Lucy can communicate—she has thoughts and concepts of her own—she just does not utter words, though she understands mine."

Roger was teaching her one sign at a time, then observing any generalization which might occur to new situations. He also was studying how the learning of one sign affected the use of another sign from the same category of signs. For example, on three different occasions she learned a sign for cry, food, and hurt. Then, sometime later, she was shown a radish. When she bit into it she signed, "Cry hurt food." After that moment of creative integration, when shown a radish she always signed either "Cry food" or "Hurt food" or "Cry hurt food." This is not an isolated or a unique example.

Roger taught her signs for food, fruit, candy, drink and cry. The signs were taught separately, on different occasions, as previously. Then, in a later session, when shown a piece of watermelon Lucy tasted it and signed "Candy fruit." When shown an onion, before Roger could teach her the ASL sign for onion, Lucy volunteered, "That cry fruit." In addition to these appropriate combinations of signs into meaningful nominative phrases Lucy, like Washoe before her, began to invent gestural signs of her own and to use them consistently.

When Lucy became almost full grown we started taking her out on a leash attached to a chain collar, so she would not play "Run away" until we got to the ranch. Quickly Lucy invented a sign for leash, a crooked forefinger held near the neck. Now she uses this sign to tell us she wants to go outside for a walk.

The signs Lucy learned gave her a conceptual and communicative tool. She now asks questions about her environment. As she sees something new she often asks "What's that?' by moving a forefinger rapidly left and right (what) and then pointing the same forefinger at the object to be identified (that). She asks this question of us, and at times of herself, as she leafs through a magazine and sees something she has not seen before. We are sure she is talking to herself or asking rhetorical questions when Jane or I are too far away for her to be talking to us, or even out of sight.

Always sensitive to our emotions, and to her own in our presence, as far as I can tell, she has used sign language to communicate feelings.

LUCY'S VOCABULARY AT AGE 9½ (after 4½ years of training)			
airplane	cry	listen	smell
baby doll	cup	look	smile
ball	dirty	make	smoke
banana	dog	me	spoon
barette	drink	mine	sorry
berry	eat	mirror	string
bird	enough ·	more	swallow
blanket	fish	no	tea
blow	flower	nut	telephone
book	food	oil	that-there
bowl	fork	on	tickle
bow tie	fruit	open	want
broom	go	orange	what
brush	grass	out	yes
candy	hammer	pants	you
can't	handkerchief	paper	yours
car	hat	pen-write	
cat	hug	pick-groom	*Names*
catch	hurry	pipe	Barbara
clean	hurt	please	Lucy
coat	in	purse	Maury
cold	key	radio	Roger
comb	kiss	ring	Steve
come-gimme	leash	rubberband	Sue S.
corn	light	run	
cow	lipstick	shoe	

One day, for example, Lucy was in the living room with Roger in the middle of her language lesson. Jane, who had been in town, drove into the driveway where she was clearly visible to Lucy through the south window. Jane parked her car and entered the house through a side door to avoid disturbing Lucy's lesson. Jane was in the rear of the house only a minute to pick up some books, then returned to her car. Lucy tried to end her lesson and go to Jane. Roger, not wanting to finish the lesson prematurely, made her sit back down and begin signing again. Lucy pulled her chair to the window and watched Jane drive away. She then signed to Roger, "Cry me, me cry."

"A Conservative Government is an
Organized Hypocrisy"

–Disraeli, 1845

If Richard M. Nixon were a psychologically sensitive scientist, instead of what he is, he might ask this question: What is the most basic function of language? His answer would no doubt be that language is not to be used for communication, but for concealment. Once, when Lucy was anxious to conceal her own mess, she used her first Freudian defense.

Shortly before a language lesson, when no one was looking, Lucy defecated in the middle of the living room floor. When Roger noticed the crime had occurred he turned to Lucy. Here is their verbatim conversation in ASL.

Roger: "What's that?"
Lucy: "Lucy not know."
Roger: "You do know. What's that?"
Lucy: "Dirty, dirty."
Roger: "Whose dirty, dirty?"
Lucy: "Sue's."
Roger: "It's not Sue's. Whose is it?"
Lucy: "Roger's!"

Roger: "No! It's not Roger's. Whose is it?"

Lucy: "Lucy dirty, dirty. Sorry Lucy."

This incident excited my imagination. Given the conceptual tool of language Lucy had told her first lie to avoid personal responsibility. She must have had a primitive concept of "good" and "bad" or she would not have lied; has Lucy taught us that "morality" is the mother of deceit? In other words, without a category of "good" and "bad"—and Fouts' research has demonstrated that Lucy can categorize—there would have been no need to lie.

Also she plays dumb; for example, she might act as though she does not understand a sign she has used correctly on many previous occasions. Or when I tell her to do something she does not want to do, she might look at me with a blank and vacuous facial expression, then turn away a moment later and ignore me.

Learning ASL has enriched her life. It gave her the power to ask questions about her environment. Before ASL it was clear that Lucy had a primitive sense of beauty and aesthetics—she would draw, use finger paints, and crayons, and work at her production until she got it just right by some unknown chimpanzee standard. And I have seen her sit and stare as though enraptured at a beautiful sunset. Since she acquired her ASL signs she does the same things, but she now asks "What's that?" Interestingly, she has never asked "why?"

Several times I have shown Lucy a magic trick my grandfather taught me when I was a child. A coin, usually a quarter or a half dollar, seems to disappear; it is hidden in the palm of one hand and seems to reappear in Lucy's ear, or my mouth, or Lucy's mouth. Lucy is fascinated with the trick. She watches the coin disappear, and tries to find it in her ear or mouth, or mine. When she gives up searching, I make it reappear in a place she has already looked. Clearly she is fascinated and mystified by the strange disappearance and bizarre reappearance of the coin. But she never asks "why?"

Similarly, I have often seen her take an empty water glass or a goblet, study it intently as though to comprehend its emptiness, then hold it over her ear pressed tightly against the side of her head. She hears the "sound of the sea" and laughs and laughs, wearing her play-face, filled with obvious delight. Then she examines the empty glass again as though unable to comprehend how such delightful sounds can be hidden in the glass, reminding me of a question Steve once baffled me with when he was four or five: Where does the music go when you turn off the radio? (Lucy often turns the television off, for example, when she wishes to sleep and Jane and I are watching a late movie; but I have never seen her turn the set on—less likely a matter of aesthetics than the absence of a wish to quiet her mind. For example, when I watch television I wish to escape my own mind, to stop the flow of my own consciousness, rest my thought processes. Lucy does not have this need.)

The same process is true of her use of language. As mentioned, language can be used to conceal as much as to communicate; every therapist can establish this easily by observation of his client's verbal defenses. Someone once speculated that man developed language five to ten million years ago when his social life became so complex it was necessary to conceal his motives. Certainly language can conceal motives—especially from ourselves.

It is possible that Lucy has not lied except on that one occasion, even though she could easily do so, because she is more self-accepting than we are. She takes herself more for granted. This does not mean she is perfect. It means she does not require herself to be perfect. Thus she has no need to defend herself from transgressions of her own values. She is scolded frequently—for toilet-training failures, for stealing a key and trying to run away, for hiding from us when its time to go to her room, for eating one bite out of each of ten apples in the fruit bowl, for breaking into the pantry and eating a jar of honey, for stealing liquor, to name but a few examples. And when we scold Lucy she often looks guilty. We can tell she feels guilty from the furtive way she sneaks around, avoiding looking at us directly, or how she develops a blank facial expression, her eyes apparently seeing nothing when we ask, "Lucy, what have you done that you shouldn't?" However, her mood is momentary because (1) she takes our love for granted and (2) she takes her self-worth for granted. There is no self-recrimination, no self-torture, no self-hatred, as our greater language facility makes us prone to develop when, as children, our self-concepts and basic values are acquired from the reaction of parents to us.

While I was writing the above Lucy was sitting on the sofa across the room, alternately watching me and leafing through a recent issue of *Playgirl* magazine. When she tired of the pictures she put the magazine down, crossed the room, stood before me on two legs wearing her play-face. This conversation ensued, Lucy speaking ASL while I replied in English.

Lucy: "Maury, tickle Lucy."

Maury: "No! I'm busy."

Lucy: "Chase Lucy."

Maury: "Not now."

Lucy: "Hug Lucy, hurry, hurry."

Maury: "In just a minute."

Lucy: (Laughing): "Hurry, hurry, hug Lucy, tickle, chase Lucy."

How could I resist?

CHAPTER EIGHT

On Incest and Oedipus

W_HEN LUCY_ was eight years old she became sexually mature and started to menstruate. Her behavior then changed dramatically, particularly her behavior toward me. I had expected that since she had not seen a chimpanzee while growing up, all her sexual interests would be directed toward me when she matured. After all, I had raised her and was the male with whom she had the most close and enduring relationship. But it did not work out this way. Apparently, ". . . the best laid plans of mice and men gang aft aglay" for chimpanzees as well as for mice, men, and those who attempt to predict them. These changes in Lucy's behavior upon reaching sexual maturity after being raised as a human being should be described in detail, as they have great theoretical significance for personality and psychotherapy and, to my knowledge, have not been recorded previously. First, however, some background about the sex life of chimpanzees in nature is needed.

When a wild female chimpanzee comes into estrus she then is ready to copulate, and indeed she will copulate at no other time. When in estrus the skin of the genital area becomes swollen and pink. It is during this period of swelling (generally midway between menstrual periods) that the female is courted and mated by the male. She then will mate with any male who appeals to her. On occasion she may even copulate with all the males in the immediate group or area. Goodall and Tutin have also observed that sometimes a pair of chimpanzees will go off "on safari" together and copulate only with one another in a chimpanzee equivalent of the human honeymoon.

All this means that in nature no chimpanzee knows his own father, either in the biological sense as the male who impregnated his mother, or in the social sense of father-protector. The infant is raised exclusively by the mother, and no chimpanzee experiences a relationship with an adult male who helps raise him. With Lucy, of course, this was not the case. She had a father who helped her mother raise her, and since she also had a human brother, she was raised with the same social structure society provides to foster the raising of human children. What effect did this have upon Lucy? I now want to describe in detail the nature of my relationship with Lucy because it changed dramatically when she reached maturity in a way which suggests a biological basis for the incest taboo and a tendency for developing oedipal problems.

On the Taboo Against Incest

All human societies have a taboo against sexual acts between mother and son or between father and daughter. Incestuous thoughts and feelings, as distinguished from incestuous actions, are less strictly tabooed.

Shortly after we adopted Lucy I began to love her without reservation. I do not remember how long it took—I would guess not more than a week or so—before I failed to make human-animal distinctions with Lucy. She was my daughter, and that was that! The fact that zoologists would classify me as *Homo sapiens* and Lucy as *Pan troglodytes* simply illustrated for me the arbitrariness of classification and the irrelevance of academic distinctions.

Lucy and I were one spirit, if not one flesh; and at times I had the experience that Lucy was as much a part of me as Steve.

While I loved them both, there were differences in the quality of my feelings for Steve and for Lucy. With Steve I sometimes felt guilty over my decisions. I felt that I was making mistakes raising him, producing childhood trauma by too much or too little discipline, for example. I feared I might leave him with psychic scars to be exorcised in some therapist's office. Lucy, on the other hand, seemed tougher and more resistant to my moods and mistakes. If I was mad enough to curse her or even to hit her she would laugh, grab my hand, and signal that we should make a game out of it. No matter what I did, her every reply seemed to say, "Daddy, do it again," while Steve would argue, sulk, or pout. She gave me unconditional acceptance, while in Steve I sensed criticism.

It was easier to see myself projected into Steve than in Lucy, since Steve was a more intimate product of myself and more like me than was Lucy. Thus my feelings toward Steve would fluctuate with my self-esteem, and I rejected him for being like me when I was mad at myself. I never experienced this process with Lucy. I also had fewer expectations for Lucy than for Steve.

I never wished that Lucy affirm me by great scientific achievement, provide me with grandchildren, or become a financial success, while I had all these fantasies about Steve. Nor did I ever experience a fear that Lucy would discover drugs, become a delinquent, or get pregnant outside of marriage. Also, with Lucy I experienced no embarrassment about sexual feelings, whereas Steve's first questions about sex embarrassed me. So I loved Lucy without any of the inhibitions, resistances, qualifications, or practical considerations which so often and so tragically reduce the free flow of love from parent to child. And Lucy loved me the same way—uncritically, without reservations, for myself alone, while Steve, with his greater capacity for verbalizing abstractions, was often openly (and correctly) critical of me as a father.

I had grown up in a family with a kind and loving, though weak, passive, and obsessional father. My mother was a strong, critical, and manipulating guilt-producer. I never felt I could please her, though I often tried to do so. Growing up with this background I carried into adulthood a need to please women, particularly strong women whom I have loved. It was a great joy to me that with Lucy I had no such problem. I needed to make no effort to please her. I needed only to be myself to receive and enjoy her love. So throughout Lucy's childhood we took our love for one another for granted.

Each day when I got home from the office Lucy would greet me with her soft, guttural "I love you" sounds; and when pubescent she would also cover my mouth with hers in a chimpanzee greeting. Then I would mix us drinks, one for Lucy, Jane, and me, and we would sit relaxing on the sofa, Lucy in my lap or Jane's. From time to time she would hug and kiss me and then taste my drink, to make sure it was the same as hers, or drink the rest of mine if she had finished hers first.

At night Lucy slept in our bedroom, while Steve had his own room as we felt more need for parental privacy where he was concerned. She had her own high-sided crib but as soon as she could climb out of it, she always ended up in bed with Jane and me. Sometimes as a special treat we would let her go to sleep in our bed. And always, no matter which side she first went to sleep on, when we awakened in the morning Lucy would be sleeping between Jane and me. There was never an exception to this if she was in the same bed with us.

Lucy witnessed the primal scene many times, an experience I did not want Steve to have. In nature, infant chimpanzees attempt to stop their mothers from having intercourse. They scream and kick and sometimes climb all over the copulating couple, or stand on the male's head stamping their feet as the pair have intercourse. (Chimpanzees always copulate in the ventral-dorsal position, except that one species of chimpanzee, the very rare and endangered dwarf chimpanzee *(Pan Paniscus)* copulates in the "mamma poppa" position much like American missionaries. Yet we have never observed Lucy to crouch and present her genitals dorsally, as a wild female chimpanzee does.)

We did not observe Lucy attempting to stop intercourse until she was two-and-a-half years old. From that age until now, intercourse has disturbed her and she always tries to stop it. If she was asleep our sex would wake her up. Curiously, she seemed to make the same interpretation of it that human children make when they witness parental intercourse: Daddy is hurting Mother, or at least an aggressive and dangerous act is taking place. I say this because Lucy tries to stop it with the same appeasement behaviors which she uses to stop or minimize non-sexual kinds of interpersonal aggression.

Lucy never got "turned on" herself at these times, as far as I could tell; at least during childhood. Now that she is an adult I am not certain about this for reasons which will shortly become clear. But as a child, when she witnessed intercourse she always behaved in whatever way would most likely terminate sex or agression. First, she tries to distract us, usually by jumping up and down on the bed, turning somersaults, spilling a bedside glass of water, or running rapidly and noisily about the room. She might then turn the television off or rapidly flick the light switch on and off. If none of these attempts to distract us succeeded, she would grab by arms or legs (never Jane's) and pull and scream with all her might.

When she was very little, her attempts to tear us apart interested and amused me. Now that she is grown, weighs 90 pounds and is five to seven times stronger than I am, she could tear us apart literally. Since I realized how strong she was and how disturbing it is for her to witness intercourse I have been inhibited in her presence. So after twenty years of marriage Jane and I moved into separate bedrooms.

Now Jane and Lucy sleep together and I sleep alone. When there is sex between Jane and me we have to be away from Lucy, usually in my room. Even then there are hazards. One time last year Lucy awakened and came into my bedroom while Jane and I were having intercourse. I lost my erection as I suddenly became aware that Lucy was slowly but firmly biting the calf of my right leg. I did not need ASL to get the message: "Stop doing that to my mother!" I was grateful to nature for the automatic physiological response which prevented my fighting with Lucy about it. However, it was some time later before I could be grateful to Lucy for teaching me that impotence can be a defense against aggression.

But during her childhood there was no friction between Lucy and me, and there was much love.

One day while Lucy was sitting in my lap we developed a body image love game. I asked her: "Where is Lucy's nose?" And after only a few repetitions, Lucy touched her nose with her forefinger. I radiated a smile and kissed her forehead and asked her: "Where is Lucy's ear?" In rapid succession Lucy learned to touch her eyes, ears, nose, and mouth whenever I asked her to. It was a wonderful game though Lucy sometimes played dumb, staring off into space and acting as if she did not know what I was talking about. After she once learned the game she often jumped into my lap and started the game herself, touching my nose or eye or mouth, and then her own, as though to illustrate the similarity between us. Sometimes when I would say, "Where is Lucy's eye?" she would touch my eye, as if she had not understood my directions, or as though we were so close there were no ego boundaries.

I took great pride in my daughter's achievements and these games filled me with delight. After playing this body image game she often would grab my hand, pull me to my feet, and beg me to chase her, suddenly letting go of my hand and scampering rapidly about the living room, leaping effortlessly over the sofa and chairs, running down the hallway, but always looking back to see that Daddy was not too far behind. And I would love to chase her, too, throughout the large living room and long corridors.

No matter what game we were playing, or if we were just sitting on the sofa or in bed together, there would be many kisses. Lucy would spontaneously kiss me and I would kiss her whenever we felt like it. With Lucy I could make a demand which is a no-win game with a human. I could say "Come kiss Daddy," and accept her kiss without feeling that, since I had asked for it, she was only complying with my demand and was not showing affection for me spontaneously. After Lucy learned the American Sign Language of the Deaf I loved to show friends how I could communicate the request to "give me a kiss" with gestures alone. Lucy would be quick to obey when she saw me signing this message to her, though I would have to "tickle-chase" her after her quick peck on my mouth, as she would suddenly turn to make a speedy getaway after begging me to chase her.

I have tried to characterize a relationship of mutual love and great intimacy between father and growing chimpanzee daughter which included close skin contact, mutual hugging, and mouth-to-mouth kisses, because these aspects of our relationship came to a dramatic end. Both the skin contact closeness and the mouth to mouth kissing ended abruptly when Lucy reached sexual maturity.

It happened first on a Sunday morning. Jane and I were sitting in the living room drinking coffee and reading the newspaper. I can clearly remember Jane's sudden, "My God, look at Lucy, she's menstruating!" I looked up from my newspaper to see blood on Lucy's bottom as she stood on all fours on the couch, Jane carefully examining her genitals. Jane urged me to look closely, and I did so. Then I found myself coming out with the standard Freudian quip: "She's not menstruating," I said, "somebody just cut her penis off!" I was immediately struck with the absurdity of this childhood Freudianism returning at such a moment. I noticed that Jane was in tears. "Lucy's a big girl now," Jane said, and suddenly I felt guilty over my coarse and thoughtless remark as I perceived Jane's sense of loss. Her baby was now grown.

We watched Lucy very carefully over the next few months to see if sexual maturity would bring corresponding changes in her personality or behavior. We were particularly interested in such phenomenon as premenstrual tension, abdominal cramping, depression, or irascibility, which many human females report as experienced in association with their menstrual periods. I noticed only a slightly increased irascibility. Lucy seemed to me otherwise unaffected by her adult femaleness. Jane, however, thought she showed some evidence of cramping. About eight to ten days later Lucy began to change in a way which was as unexpected as it was dramatic. She started to avoid me.

At first it was a gradually reduced contact that we noticed. She simply seemed preoccupied and did not want to play with me. She sometimes would start to play with me—chase, or wrestling, say—and then abruptly stop and walk away. This had never happened before as it always had been I who ended the play. Or Lucy would be sitting on the couch and when I sat down next to her she would get up and sit elsewhere, without a glance at me or any other gesture of recognition.

Jane and I puzzled over this strange behavior but could not understand it. Lucy's reactions to Jane had not changed, and I did not feel that my attitudes toward Lucy had changed. Then the pattern of behavior and its meaning became unmistakably clear.

Fourteen or fifteen days after the beginning of menstruation Lucy's genitals became enlarged. The external genitalia increased to five or six times their normal size and became a deeper pink color from the engorged blood; her bottom became like a large rounded pink soup bowl.

This swelling of her external genitalia, of course, represented sexual receptivity. It signified Lucy was in estrus and was capable of coitus and impregnation. At that time, with her genitals enlarged, when she desired copulation and had ova available for union with sperm, *Lucy totally rejected me. She would not hug me. She would not kiss me. She would not play with me. When I tried to cuddle, hug, or kiss her, she moved away from me and would not allow any touching. Yet during the same period she made the most blatant and obvious sexual invitations to other men.* With any man but me, even with a stranger or with a man she did not usually like, she would jump into his arms, cover his mouth with hers and rub and thrust her enlarged genitals against his body.

Lucy was eight years and three months old when this pattern first occurred. In the ten months between then and the time of this writing we have observed the same sequence each month; when her genitals are enlarged she will permit no close physical contact with me of any affectional nature. She will not hug, kiss, caress, or cuddle with me, but uninhibitedly demands all these behaviors from any available man, which has been disconcerting to Fuller Brush men, Bible salesmen, and census takers who have knocked on our door. During the remainder of the month, when she cannot get pregnant, she will kiss and caress me as usual.

These observations were completely surprising to us (and somewhat disappointing to me). I had thought that Lucy would be "imprinted" on me and direct her sexuality toward me when she reached sexual maturity. I had even had fantasies of copulating with Lucy and had cracked jokes about it, teasing Jane about how our daughter would be a perfect subject for an experiment in cross-species sexuality, and that no marital infidelity would be involved since Lucy was "in the family" and a chimpanzee anyway.

But sexual imprinting did not occur in the predicted fashion, and the redirection of her sexuality toward men outside the family suggests an incest taboo, at least in female chimpanzees. Since the taboo correlates perfectly with her capacity to conceive, the determinants of the taboo must be biological, probably endocrinological. This taboo extends to Steve as well. When they were children together there was much free-flowing affection between them, which has continued to this day—except when Lucy is fertile. At these times she also avoids Steve.

One time Jane Goodall visited Lucy and stayed at our house overnight. I mentioned these observations about Lucy's incest taboo to her. Dr. Goodall said that she had observed this same kind of behavior during her studies of free-living chimpanzees. The young chimpanzees that Dr. Goodall observed in the Gombe Stream Preserve of Tanzania often "play copulated" while growing up together in the same peer group. However, as soon as the females began to ovulate they would cease copulating with childhood playmates, leave the group in which they had been raised, and join strange groups so that they bred with unrelated males. Dr. Goodall said that she did not understand what biological mechanisms produced this behavior. Nonetheless, the advantages to the species are obvious. The outbreeding increases the variability of the gene pool and makes fortuitous combinations of genes more likely. Thus, though the biological mechanisms by which the taboo is created are unknown, they are of great importance.

Incidentally, I found Jane Goodall to be a woman of great dignity and charm. And Lucy took to her immediately. From across the room Lucy recognized immediately a person whose attitudes towards her were totally positive, and she ran towards her making her sounds of warm love greeting and jumped into her arms and kissed her. Dr. Goodall was such a self-actualizing person, in Maslow's sense of that term, that I felt and Lucy acted humble in her presence. Whether she was in our home or lecturing to two thousand students I never noticed one bit of anxiety or pretense. She stayed one night with us while lecturing at the University and though heavy demands were made upon her by Lucy, Jane, and me, the two thousand students who came to hear her, faculty wives, and bureaucrats, I never saw a moment's departure from an attitude of relaxed, warm cordiality towards humans and chimpanzees as well. Nor did she ever show any evidence of fatigue even though she lectured for several hours and we kept her talking until very late.

At a cocktail party following her lecture a visiting New York psychoanalyst whose sensitivity was soluble in alcohol said: "Dr. Goodall, now let me ask you the $64 question. In all the time that you were out in the bush alone with the chimpanzees, did you ever get sexually turned on by them?" Without a second's delay and with no change in inflection or expression Dr. Goodall replied: "No, I never observed that phenomenon in myself, but I saw something quite like it in a female graduate student. It was often very hot and several female students were fond of working without wearing their shirts. But there was one female graduate student who always put her shirt back on whenever the chimps entered the camp."

I found this a very interesting observation as throughout the years of my deep affection for Lucy I never experienced sexual desire for her. I had had fantasies of humorous situations which might occur if Lucy made sexual advances to me, or of artificually impregnating her with my semen, even though I knew this would not work. But these were thought process; and at the level of feeling I never experienced a conscious sexual desire for her.

Indeed, it is curious that I never experienced desire for her even though Lucy would see me nude and attempt to put my penis in her mouth. Although this may appear sexual to the reader, it never felt that way to me. I always felt that her mouthing my penis was exploratory rather than lustful as she never subsequently "presented" her genitalia and she never once attempted this behavior when she was in estrus. Furthermore, Lucy attempts to mouth my penis whenever she sees it, whether I am urinating, bathing, or have an erection. As a matter of fact

I think it is accurate to say that Lucy is fascinated by the human penis since she attempts to explore it with her mouth whenever she can, unless it is mine and she is swollen in estrus. In that case she avoids me entirely, though any other human penis which might be available at that time is a source of fascination to her.

About the time Lucy started coming into estrus every thirty days I noticed that Steve was collecting *Playboy* magazines. The housekeeper would leave them on the coffee table in the living room and Lucy and I would thumb through them. Lucy had no particular reaction that I could see, or at least no more than she had when she looked at the pictures in other magazines. But I reacted to them. I remembered that when I first reached sexual maturity I had enjoyed masturbation while looking at pictures of nude women, and in that recollection an experiment was born.

The next time I was in town I entered an "Open 24 Hours" grocery and magazine store and bought a copy of *Playgirl* magazine, with a feeling of gratitude to the Women's Liberation Movement for making the experiment possible. I leafed through the magazine as the clerk prepared my change. "If that's what you're interested in," he volunteered, handing me my change, "you oughta see *Cosmo* this week. It's got a nude fold-out of Burt Reynolds."

"Never mind, this will do," I said as I noticed the penis was fully exposed in the *Playgirl* pictures.

Lucy was in full estrus, sitting on the sofa having a gin and tonic with Jane when I arrived home. It was the perfect time and place for the experiment.

"For Lucy," I said, handing her the copy of *Playgirl*. She accepted the gift nonchalantly at first, then became increasingly excited as she leafed through the pages. As she came to each picture of a nude male, her excitement visibly increased. She stared at the penis and made sounds similar to those she utters when looking at some delicious morsel, a low, gutteral ". . . uh, uh, uh, uh" sound. She stroked the penis with her forefinger, cautiously at first and then more rapidly. On some pictures she would first stroke the penis with her forefinger, get very excited, and then mutilate it by scratching it with her fingernail. When she finished with one picture she would turn the page and start on another, alternately caressing or scratching the penis. She completely ignored Jane and me and was totally absorbed in her experience as she worked her way through the magazine. She did not caress or scratch any other part of the photograph—just the penis.

When she came to the centerfold she carefully unfolded it, studied it for a moment or two, then got off the sofa and spread the large slick photograph of an aspiring young actor on the floor. She stood on two legs over the photograph and carefully positioned herself to lower her pink and swollen genitals onto the penis. She rubbed her vulva back and forth on the penis for about 15-20 seconds, maintaining contact with the photograph. Then she changed her movements and started bouncing up and down "on the penis." The bouncing movements were of even shorter duration, approximately 10 to 15 seconds, certainly no more, though Jane and I were too fascinated to tear our eyes away and look at a watch. She suddenly ceased bouncing, stood up, and walked about the room. I could not tell what she was feeling, but I thought the experiment was over when she seemed to pause in mid-stride, her face assumed an expression I labeled "thoughtful" and she suddenly returned to the photograph. Again she positioned herself over the penis and slowly lowered herself onto it. This time, however, instead of making contact with it she squatted above it, and with her vulva no more than an inch above the penis, she urinated on it, carefully controlling the trickle of urine so that it landed directly on the penis. I said, "My God, Jane, Lucy's pissed off," and I wondered if I had not discovered the ultimate primordial origins of that phrase. In any event I realized I had never received such a unique response to a gift of love.

On the Oedipus Complex

Lucy's incest taboo during estrus suggests a predisposition for developing oedipal problems if the environment should present certain stresses. Consider, for example, what might happen if I were to press Lucy for sexual closeness—for example, to try to kiss her when she is most anxious to avoid me. Lucy would experience a conflict between her biologically-determined desires to avoid me and seductive social forces—one cause of neuroses in humans. And that conflict is exactly what happens. To test this idea I have on many occasions pushed my attentions on Lucy, trying to hold, cuddle and kiss her during fertile periods, always with the same results: Lucy tries to get away then, when I hold her, she looks at me and screams in terror. It is the same sound and the same lips-turned-back, gums-and-teeth-exposed face that I identify as terror inspired by her own aggression. In this situation, her aggression has been mobilized in the presence of forces which inhibit attack: the presence of family members. While screaming—whether in terror, or rage, or whatever it might be—she then lunges at me and I immediately retreat, experiencing the fear that I would have been bitten had I not done so. Notice that Lucy could attack and make me withdraw rather than internalize the conflict-inspired aggression and become neurotically depressed, as human children often do.

Lucy's incest taboo was fascinating to me when I first became aware of it. I had, of course, known of Freud's concept of an instinctual incest taboo since I began reading him in high school, and my first introduction to psychotherapy on both sides of the couch had been Freudian. But more important than that, I have had much experience with *psychological* incest and, unprotected by a similar taboo, I discovered how destructive it can be.

Although my experience with incest was psychological rather than physical, psychological processes are usually correlative to physical ones. (I do not mean to imply a

Lucy enjoys looking at pictures—sometimes with Jane, sometimes alone. The nude centerfold in "Playgirl" magazine brought an unexpected sexual response to this visual stimulation. In the picture opposite, Lucy's kisses were directed to the photograph of the penis. Below, on the right, "she stood on two legs over the photograph and carefully positioned herself to lower her pink and swollen genitals onto the penis." In the wild, chimpanzees have been observed to copulate only in the dorsal position. Lucy has apparently learned the ventral position from observing her human parents.

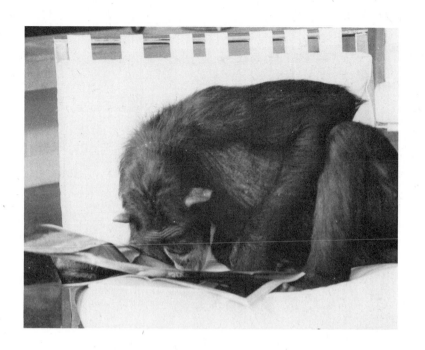

psycho-physical parallelism, or to resurrect the mind-body problem. I think physical and psychological processes are different aspects, differently labeled, of global organismic functioning. I am a monist. I am my mind, and I am my body. But since Decartes divided the world into those aspects which had weight and occupied space and those which did not, our language developed in ways which make it difficult to talk in any but dualistic terms. Living for so long with such an integrated creature as Lucy, who has not read Decartes and whose mind and body are so clearly one, dramatized for me the unity of those processes we label psychological *and* physical.)

I once had gone into psychotherapy with a therapist who promoted incest by encouraging his clients to enter into many roles and relationships with him in addition to the therapeutic one; and he failed to set limits on any dual roles or relationships a client might propose. Little has been written about psychological incest. However, all human societies prohibit physical incest. We have known since Freud that wishes to commit incest are not unusual, though acting on them is fairly rare, though certainly not unknown, and always destructive. Psychological incest can also be destructive because, when everything of emotional import is kept within the family, the participants are not learning roles and relationships which are effective outside the family. Incestuous therapeutic relationships are destructive to the therapist as well, and when I finally terminated with a feeling I had wasted many years, I felt that my therapist had become very bitter and depressed. He experienced acute anxiety in any situation which was not centered about him, or which he did not control, and I thought he was incapable of an enduring relationship with anyone who was not dependent upon him. These either were his reasons for "Keeping it in the family" in the first place, or constant incest had reduced his capacity to function outside the family. Since there seem to be no built-in defenses against incestuous relationships in man, Lucy's taboo against incest must be another example of what Walter B. Cannon has called "the biological wisdom of the body."

The subject of psychological incest brings us to a profound experience that I had during a stay at Esalen, Big Sur, California. I realize that *Growing Up Human* is basically Lucy's story. But in a very real way, it is also my story, as I had a good deal of growing up to do as a human myself. I insert it in the following chapter, though I know I risk interrupting the flow of the narrative about Lucy. I ask your indulgence and understanding, as the experience affected me and my life with Lucy and the other members of my family in ways I could not possibly have ever foreseen.

So, let me talk about myself for a chapter.

CHAPTER NINE

Keeping It in the Family

"There is no Birth of Consciousness without pain."
—*Carl Gustav Jung*

WILLIAM JAMES was my first God, and then came Freud. But by the time I reached graduate school, I was an admirer of Wilhelm Reich. I thought his *Character Analysis* was a masterpiece, and that psychotherapy as a process of personality change had to be based on the ideas of *Character Analysis*. If I were to describe myself as a therapist before I went to Esalen for the first time, it would be as a pre-orgone Reichian. (Reich lost me when he "discovered" orgone.) I had been touched by the humanistic psychology movement, which seemed a healthy antidote to the sterility of Behaviorism, and I had even published a paper on "Free Choice and Personal Responsibility in Humanistic Psychotherapy." But that was at the level of theory. In practice I stayed within the general framework of *Character Analysis*, and I was a conventional therapist. I stuck to "official" technique and theory, clinging to that with which I was most familiar, not yet trusting myself enough to use my own clinical judgment and to modify what I did in terms of "what seems best" with the particular person at the particular moment.

I went to Esalen, at Big Sur, California, for the first time after thirteen years of individual psychoanalysis. My psychotherapy had been a series of disappointments for the most part. I had first spent three years on the couch of an orthodox psychoanalyst, a training analyst, when I began graduate school. After I got my Ph.D. and had had a few years of practice, I again entered therapy. My therapist this time had no formal training in psychotherapy beyond the Ph.D. but he had a great reputation and enormous charisma. When I first met him I was enormously charmed, and I soon came to love and trust him without reservation. In psychotherapy with him I continued to grow as a person, of course. The neurotic symptoms of my youth gradually disappeared. Before therapy I had flunked out of U.C.L.A.; shortly after starting it I began to excel academically. Honor rolls and honor societies, a Ph.D. and a departmental chairmanship came in rapid succession. I was even reasonably happy at the time, but I felt there must be more to life than being "reasonably happy" and "reasonably successful."

My friends and colleagues treated me with dignity and respect and expressed admiration for my ability as a therapist, teacher, administrator, or consultant. However, I secretly felt like a fraud and that their praise was undeserved because they saw only my external appearance, not the "real me." I felt I was too intellectualistic and substituted words and concepts for emotional substance, led both by neurosis and my profession as a psychologist to abstract myself; to be an observer of myself rather than a full and active participant in dynamic interaction with the world. I often was in my words and concepts, not behind them. I was afraid of being myself in intimate relationships with others, particularly if it might mean a hostile disagreement or a confrontation between myself and someone I loved because I felt that no one could possibly like me if they really knew my feelings. In other words I felt that what other people saw when they observed me was a mask rather than my true self and yet, paradoxically, I felt that my self-esteem depended precariously on maintaining the mask even though the mask was not me.

My therapist called me a "narcissistic character disorder" for being concerned with the mask of appearance which I felt compelled to wear, and I am sure he was applying the correct label. But I have never believed the pontifical application of diagnostic labels was either helpful therapeutically or valid scientifically. However, I continued to love and to trust my therapist, often accepting uncritically the most pontifical of statements, while at the same time I sensed that I could grow no further with him. But I could not admit it, as I was blinded by incest and transference, not having Lucy's automatic avoidance mechanisms.

We gradually had become involved in many intricate dual relationships besides the therapeutic one. Our administrative and financial arrangements had become very complex. When I needed a home he sold me one—at no down payment and no interest, which I felt exemplified his benevolence, for he behaved this way with other grateful clients. It was years before I realized how I was being controlled through obligation and guilt, and that his "benevolence" redirected hostility against myself. We socialized and traveled together, attending professional meetings both at home and abroad. We became partners in many joint business ventures so that the therapeutic relationship was no longer clean.

I had observed many times that my own clients progressed more rapidly and made more pervasive changes working therapeutically with me than I was making with him. I often expressed the feeling that it was our multiple relationships outside the therapeutic hour which prevented my growth. We both knew, of course, that by the ethics of psychology, psychiatry, and psychoanalysis the dual relationships that we had were unethical, and I felt the psychotherapy profession had defined them as unethical for good reasons. He replied that there was "a higher loyalty" to the therapeutic process itself, not to the "understanding of psychotherapy institutionalized by the bureaucrats of our profession," and that as long as he was not exploiting me, our therapeutic relationship was ethical in a broader sense. I did not then see how I was conspiring in my own destruction when I attributed a benevolent omnipotence to him, and great weakness and fallibility to myself. I therefore believed him when he said our multiple extra-therapeutic relationship would help me as I ". . . lacked a capacity for intimacy," and could "grow by identification" with him. Looking back on those tragic years and trying to understand them, I believe he did his best, and I know I did; but he was unaware of his needs for power and control over others to maintain his self-esteem, and I was unaware of my need for a God to avoid the anxieties of autonomous selfhood.

The psychotherapeutic process is vulnerable to several kinds of insidious corruption. Every time I said I was making no progress and was considering termination, he raised the spectre of "acting out." He said I was running from some painful unconscious aspect of myself rather than exploring it verbally and analytically. So to avoid "acting out," and to take advantage of what I considered his superior knowledge and wisdom, I stopped making any choices or decisions unless I first discussed them with him. And at that moment I stopped growing as a person! I know now that one grows only by making his own conscious choices and decisions. Then, if they work out, self-esteem is enhanced; while if one's decisions lead to disappointment or failure one can see the consequences of his own actions and change them in the future. Growth and learning are possible in either case, as long as the choices and decisions are truly one's own. But the responsibility for my choices and decisions was shared with my therapist; as such they were neither mine nor his entirely, but a product of two closely interrelated mentalities. I frequently could not decide whether a particular idea came from my mind or his, and I was very confused.

Paradoxically, I suffered from both too much and too little courage. I had too little courage (and too much guilt) to terminate and risk disaster to our joint business and financial ventures, including the mortgage on my house, and I had too much courage to terminate when I might be running away from myself. So instead of deciding to leave and doing so, I stayed and published papers on free choice and decision-making in psychotherapy, conceptualizing neurosis as an inability to choose and to decide. And when my conscious experience of myself and my therapist differed drastically from his experience of himself and me, I based my actions on *his* experience, not my own—and published a series of papers on suggestion effects in diagnosis. How blind I was! I even published an experiment in which psychiatrists and psychologists had diagnosed a perfectly normal, healthy and effective man as neurotic or psychotic when they interviewed him after hearing an authority figure casually opine that he looked neurotic or psychotic.

It is indeed tragic that such a noble invention as psychotherapy can be so easily corrputed. The lack of self-esteem for which I had sought psychotherapy in the first place led me to overvalue my therapist, to transfer to him an illusion of perfection and greatness, and to deny my own experience in favor of what I was told by a charismatic therapist. I rejected my experience *as given* in favor of abstractions from my experience, for example his hypotheses about my unconscious. Conversely, his low self-esteem led him to accept the role, to erect a facade of pontificating omnipotence, and to become advisor and manipulator of his clients. But I was acting out, not analyzing the transference and could not recognize the corruption and inherent destructiveness of the incestuous extra-therapeutic relationships which made him the center of my life and prevented me from discovering that the power and wisdom I perceived in him were projections. It took me many years to learn that there is no power and wisdom for anyone which is not illusory except that which he can discover in himself. Nonetheless, an illusion of enormous power and vanity may accrue to the therapist who believes such projections and allows such relationships. Illusory power corrupts as absolutely as absolute power. Narcissism and meglomania are occupational hazards of psychotherapy.

Self-esteem is a precious but fragile gift. Without it we despise ourselves and live neurotic or robotized lives, following the rules of others to avoid the experience of ourselves. But self-esteem is easily perverted in the other direction. The therapist is vulnerable to seeing himself as a superior being deserving of his clients praise when they do not recognize their own strength and power. If this happens the therapist may become a leader, an expert administrator and manipulator of the people who pay him tribute rather than a fellow pilgrim in the exploration of the complexities and paradoxes of human experience. The therapist so corrupted becomes addicted to praise and flattery, then must depreciate his clients as dependent, weak, or inadequate to maintain his sense of superiority, and the relationship becomes profoundly destructive. These are some of the meanings I see in the Zen phrase, "If you meet the Buddha on the road, kill him" which Sheldon Kopp used as the title of his book on psychotherapy. Each man must find his own way, and we therapists can only ask the right questions at the right time, and facilitate as a catalyst, reflect as a mirror, or share as a fellow human being.

One day my therapist suggested that I go to Esalen and be in an encounter group. I was terrified. (I later understood that he had suggested Esalen to me so that he might experience it vicariously; he had never been there himself, or in any group except as a leader. I also believe he could take no responsibility for ending our toxic relationship and that he had resorted to a referral to Esalen rather than making use of personal therapy himself, professional consultants, or a simple, honest "We are not getting anywhere anymore." For example, a friend of mine who was one of his other long-term clients said, "I have been in therapy with you for twenty years and look at me; I have the same anxieties that I had when I started."

With a warm smile and no hesitation he replied, "Contemplate what you might have been like without twenty years of therapy.")

So with enormous fear and despair I went to Esalen for the first time, to participate in a one-week encounter group led by Seymour Carter. It facilitated the most powerful and rewarding experience of my life, and changed my personality and my life overnight.

To see Esalen for the first time is to be overwhelmed by its sheer physical magnificence. The sea meets the earth in a swirling clash of waves against rocky cliffs. Color is everywhere, in sunlight flashing off the water, in brilliantly colored flowers, filtering through the clouds rising above the ocean, sometimes blinding against eyes which hesitate to blink lest some form or color be missed. Flowers and trees grow out of rocky crevices, sometimes parallel with the turbulent surf below. Buildings perch precariously two hundred feet above water which manages to stay clear in spite of the turbulence of the surf. Sea otters play about gigantic boulders, and occasionally a whale may be seen blowing in the distance.

The people are as spectacular as the physical surroundings. At that time I was a middle-aged, hyperorganized, super-intellectualistic, conventional professorial type—a bureaucrat, too. I had just been forced out as chairman of a very conservative psychology department at a very conservative university. I felt hideously out of place among people who followed no pattern. Every manner of dress and undress was in evidence. Women wore long hair, muumus, jeans, dashikis or nothing. Occasionally I could see a nude nymphet running towards a swimming pool in the distance. There were men in facial paint, headbands, boots, sandals, or barefoot. One wore a hunting knife about his waist as he sat shirtless, looking at the ocean, playing bongo drums. I wore the only business suit in sight, and the world at that moment seemed crazy. Confused, disoriented, and wildly depressed, I experienced great anxieties as I registered, went to my room and waited for the first session of the group.

For the next several days I participated in what I initially considered organized madness. I think that psychologists have a tendency to label psychotic anything they do not understand, or which follows unconventional rules. At least at that moment I did so. In the group whenever I asked a question, the most respectable activity in academia, I was asked to change it into a statement. When I made a statement, I was asked where I felt it in my body—which seemed absurd to me since I didn't feel it in my body at all, but thought it in my head. But I had committed myself to full participation, to following all the rules (or better, lack of rules). As I expressed how I felt, what I experienced, staying in the immediate present, and interacting with what everyone else reported, I became deeply involved. My conventional past faded into the background. The therapeutic-like activities in which I engaged were as different from the psychotherapy with which I was familiar as day from night. I talked to my wife—not about her—but to her, in an empty chair. I told the group that I had come to Esalen after many years of individual therapy, and still felt incomplete; I was too much a disembodied intellect, unable to lower my mask, be real, experience negative confrontations. But that was all in the past. Now, there was only the present. And I did everything I could to stay with what I was experiencing at the moment, and to let the past go. I still don't know for sure how it happened, but in the madness of the moment, and with the feeling that in that counter-cultural world where I was not known, anything was permissible, I got carried away with hostility toward a young man in the group. Seymour asked me to wrestle him and I did so, without thinking about it in advance or "observing" myself wrestling him. Even though he was thirty years younger than I, I beat him. He was much stronger than I was but I was totally involved in the moment, carried away by rage. Functioning as a whole, without inhibiting myself enormously, increased my strength. It was precisely at that point that I began to lose my fear of aggression and to realize that hostile and aggressive feelings could be the source of great strength and creativity if I could only express, accept and integrate them.

When the group broke up that night I went to my room rather than to the communal hot baths, as was our custom. I wanted to be alone, to try and absorb the experiences I was having. Suddenly I began to sob, and then to cry in earnest—deep sobs that convulsed my whole body. I had not yet learned to "go with" tears, not to fight them. I had been brought up on "Big boys don't cry," and at first I tried to control my tears. But I soon learned that crying is nature's way of organismically regulating emotions and as I realized what I was crying about, I no longer fought to control my tears and I became a mass of pulsating, sobbing grief.

I cried all night long over the loss of my son, Charlie Brown. When Charlie Brown died I had really not been able to work through the grief, because I could not tolerate the intense pain his death meant for me, because of the "Big boys don't cry" madness with which this culture infects us. So I had told myself that Charlie was a chimpanzee, not a person, and I had suffered only the loss of a beloved pet—not a son. So I still had "grief work" to do before I could completely let go the painful memories of Charlie Brown. But at Esalen, in a new and foreign world, disorganized by the group experience, I could surrender to the tears. I cried all night long. When dawn came I had been asleep only momentarily, if at all. I was at the baths as the first light permitted a view of the mountains through the fog. I was filled with disturbing feelings I could not understand. Shortly after dawn, several other people came to the baths. I watched two of them do something that I had never seen before. The woman lay on her back, nude, eyes closed, supported in the warm water on the palms of a man who rocked her back and forth as he hummed a low umumum. The whole scene seemed incredibly peaceful, and I began to crave it. I asked the man, whom I did not know, (but I had already learned that at Esalen one does not need an introduction) if he would rock me in this fashion. He said he would when he and the woman finished, and he continued to gently move her back and forth on the surface of the warm water, and to hum to her.

I again found myself having very strange feelings, as though I could no longer grasp what was happening to me, as though I could no longer control the flux of my own experience. My perceptions were beginning to change. I was feeling sensations in my body which I could not understand. They were neither pleasant nor unpleasant, just radically different bodily sensations from anything I had felt before.

When my turn came and I was held and gently moved through the water, I closed my eyes and immediately saw an eidetic image of my mother's face as a young woman. I then had a fantasy of being born again, and it was a beautiful and peaceful fantasy. But suddenly the tranquility of the moment was destroyed by a series of thoughts which intruded themselves into my consciousness. They were thoughts and feelings of embarrassment over having just been fired as chairman of a psychology department. The stranger (I later found out that he was a Gestalt therapist named Koleman) could see that I was preoccupied and unable to let go and be with the rocking. I started to tell him about it, and he asked me instead to sit on the edge of the bath and talk directly to the person I was thinking of. It was the then-president of the university, who had fired me as Chairman of the Psychology Department. He had done so in the best tradition of Nixonian sanctimony, himself taking no responsibility for the action, by asking me for an undated letter of resignation while promising not to accept it. I had written the letter of resignation and he had promptly accepted it.

I talked directly to him, telling him how angry I was for such fraudulent and dishonest actions. I also expressed the embarrassment I had felt over being fired for the incredibly ignoble reason of shooting a pig. Actually, several graduate students had shot a pig on my land while hunting quail. The pig had belonged to a neighbor, and I had covered up for the students. The irate neighbor, whose pigs were always trespassing on my land, had told a rural legislator who then had complained to the president of the university about it. The president used it as an excuse to force me to resign as chairman, though I stayed on as a professor. His real reason had been that I was supporting a Ph.D. program in clinical psychology in the midst of a war between clinical and academic psychologists. The civil war in the psychology department had become very noisy, as we had been training psychologists as psychotherapists, not scientists, an unpardonable sin in doctoral programs accredited by the American Psychological Association. Nonetheless, absurd as this series of events was, I felt angry, embarrassed, and ashamed about the episode, and I expressed these feelings with great intensity. As I spoke directly to the president, my feelings began to change. I could even see humor in the incident: Distinguished Jewish professor of psychology fired for shooting a pig—the totem animal of antiquity. I suddenly realized I no longer had any secrets. I needed no defenses, no masks. I had nothing to hide. Moments later the experience began to happen.

The experience was so powerful that as I write about it nine years later I can still feel my flesh crawl with a sense of awe and my eyes fill with tears. I began to see light everywhere. The sun was not yet up and I knew that it was not sunlight that I was seeing. Yet all around me, the mountains, the sea, the waves, the people were becoming very bright, illuminated with an extremely intense and bright light. At the same time I began to feel an exquisite pleasure, a pleasure greater than anything I had ever felt before. I had never had an orgasm that was either so powerful or so pleasurable or which lasted so long. The pleasure was so great and the light so bright and intense, and together the whole experience was so strange and powerful, that I started to run from it, to maintain control over my consciousness, and I might have done so had I not heard Koleman say, "Let it happen."

I was trembling so much that I could hardly walk. He helped me to a wooden railing where I stood, nude, at the edge of the cliffs above the sea. I looked down and watched the waves hit the gigantic boulders at the base of the cliff. As I stared at the waves I suddenly became aware of a feeling which is incredibly difficult to describe as I write about it in retrospect with my organized and controlled consciousness. But at that time, as I stared at the sea, I could feel the energy and force of each wave flow into me. As the waves hit the cliff, their energy came into me, because I in some way experienced myself as having become the cliff itself. As each wave crashed against me I began to feel a strength I had never felt before. I continued to watch the waves, opening myself up to receive their energy, as they hit the cliffs with which I had merged. Sometime later there came another change. At first I had been receiving the energy of the waves; at that moment there was a subject-object split in my perceptions. I, the subject, felt the energy of the waves, the object. Gradually the split disappeared and I lost all physical and psychological boundaries; there were no more subject-object distinctions. I *became* the waves and I experienced the turbulence and the power of the tides as we crashed against the boulders, individual consciousness dissolved in the violent sea. I do not know how long that experience lasted; waves are timeless, eternal.

As I slowly came back into myself I looked to the south where two green mountains sloped to meet the sea. As I watched the mountains I began to experience a completely different way of seeing them. They started to glow with an intense, eerie light, and I was astounded to realize that I could see directly into the mountain. As I looked into the

mountain, to my further surprise, it was not only strata of
rock that I saw, but moving about through the strata were
images from my childhood. I saw myself as a child, with
my mother and father, and other children whom I had
known as a child—forty years earlier. Suddenly the images began to act out incidents and scenes from my
childhood in a way which made my life seem meaningful
and purposeful for the first time. As I write this I know
that the images must have been in my head, and that the
mountain was out there, apart from me. At the time,
however, I did not experience it this way. The images
were of people I saw inside the mountain. They would
sometimes come back to the surface, then go in again. But
I saw them both inside the mountain and on its surface
and did not experience them as in my head. I watched as
they acted out incidents and early traumatic experiences,
while I viewed the reenactment with a tranquility I had not
had in childhood. I saw pictures of myself as I was then,
and as I now am, blending into the mountain and becoming one with nature, a feeling that I had never had before,
and that my life now had meaning and purpose. The
bright light continued as did the exquisite pleasure,
though the pleasure gradually became a glow of satisfaction rather than a brightness so ecstatic that I felt I could
not stand more of it.

I believe that I rocked back and forth on the balls of my
feet, then to my heels, as I stood there, naked, alternating
between the view inside the mountain and going back
periodically to become and absorb the energy of the
waves. I was completely unaware of the other people at
the baths throughout most of this experience. As I think
back on it, though, I may have noticed one or two people
standing nearby at the beginning, as I can remember the
feeling that others must know how radiantly happy and
energized I was. Later, I found that this had been the case
because several people came up to me, people I did not
know, and threw their arms about me and embraced me.
They said that they had been grateful to be around at that

time because they could see that something very powerful and wonderful was happening to me.

Gradually the experience came to a close. The light slowly faded and I could see into the mountains no more. The waves returned to being just waves. I was spent and exhausted. I felt a great humility, because I had known a completely different way of experiencing the world. I was grateful—to nature, to whatever Gods there may be, and to myself, to know that I was capable of having this kind of experience.

When my group met again later that day I found I did not want to talk about the experience. I thought that talking about it might "talk it away" through the processes of intellectualization, and I felt myself still changing and did not want to interfere, as talking can sometimes interfere with any intense emotional experience. Too, I felt hopeless at the prospect of attempting to communicate the experience. Words are such hollow conveyors of emotionalized meaning unless one is a poet, and who could understand without having had such an experience himself? So I told no one about it. It nonetheless was clear to the group that I had changed dramatically as a person.

As the week drew to a close Seymour asked us to stand in a circle, our arms about one another's shoulders, and to go on a fantasy trip inside ourselves. When I closed my eyes and opened myself to whatever wanted to come into awareness, I saw myself in the center of a long cylindrical tunnel. Standing there I could look either to the right or to the left. At one end of the tunnel I saw a bright light—the same bright orange light I had seen earlier—which I interpreted to mean continuing life. At the other end of the tunnel I saw death. I did not move through the cylinder in either direction for it made no difference which way I might go; I felt both were natural parts of one another, different sides of the same coin. I have never again feared death.

Since that week at Esalen I have been a different person. I lost my mask; and since then I have experienced no difference between myself as an identity and my appearance when I observe myself. I am I and I am me. I am now behind my words. I retained the capacity to let go controls over my stream of consciousness, no matter what it brings to my awareness; I control my words and actions, but not my experience. I retained the capacity to allow the free flow of tears and to explode into grief or rage or laughter, depending upon the situation. My attitudes toward other people remain dramatically changed. I lost a suspicious, administrator's attitude toward others, and acquired instead a feeling of closeness and warmth towards people which has persisted to this day—nine years later. I now see other people as wonderously complicated beings, potentially capable of beautiful and powerful experiences—but my hero worship was cured.

I had not yet terminated with my "there's nothing new since Freud" therapist, so when I returned I told him about the experience. He smiled and said, "Yes, yes, of course! Psychotic ecstasy, brought about by the lack of external structure and familiar reference points. It may take you a while to come down to reality."

I do not care whether that experience is labeled "psychotic" or not; the long-haired, jeans-clad secretary who typed the first draft of this chapter said, "Wow! That must have been great acid!" and she at first did not believe that I had not used psychedelics. Both comments are in the same category—reducing the threat of the unknown and uncontrolled by applying familiar concepts and labels, though even so the secretary, who had tripped on LSD, was closer to her capacity for mystical or peak experiences than the therapist who simply sneered and patronized.

When I looked into the mountain the images acted out many scenes from my life. One of them is worth describing here because it helped me understand why I had remained so long in a toxic therapeutic relationship. The scene also illustrates one difference between Lucy and human beings and a critical problem of human existence.

The picture I saw was of five children—Henry, Ralph, Elmer, David, and me—sitting in a semicircle on the floor of a Temple. Judging from my size I was four or five years old at the time, and the Sunday school teacher, a Mrs. Hertzmark, was saying something I had not heard before: I was a Jew. "What's that?" I asked. "The chosen people of God," she replied. "What's God?" I asked, becoming very confused. She answered that God was "The creator of the heavens and the earth" and she then gave a short history of the Jews. Ralph, Elmer, Henry, and David were frozen in rapt attention. I could see myself squirming. "If the Jews were the chosen people of God, why were they suffering so much and why were they enslaved?" I asked. "The ways of the Lord are mysterious and beyond under-standing," the image replied. I didn't like that at all. "I don't want to be a Jew," I said. She silenced me with "You can't help it, you are one, and you should be proud of it," and continued with her history lesson. I listened silently, but I was very uncomfortable. There was much going on inside me, some of which I still do not understand. By the time she reached the part about Moses—after whom I had been named—leading the Jews from slavery under Pharaoh in Egypt to the Promised Land I had reached a decision which was to affect me profoundly throughout my entire life, and which almost destroyed me. She was telling about the waters of the Red Sea dividing to let the Jews pass over, then coming together to drown the pursu-ing army when I screamed it aloud: "This is not true! I don't believe it! There is no such thing as God!" I screamed and yelled and kicked and cursed. In one sense I was having a temper tantrum, but in another sense I was having a religious experience. But it was a conversion to atheism! Then the images had faded away and I remem-bered a terrible feeling of being hideously different from other people. I was not a Jew. I was not a Christian. I was nothing—except, perhaps, a monster for being so different and not believing what my friends and parents believed so devoutly. I was totally alone—unidentified with anyone or anything, aimlessly drifting in a mysterious and threatening universe I could sense but not comprehend. My parents were no support, for they were Jews and believed in God—and I knew with the emotional intensity of full religious conviction that there was no God.

After that experience my childhood was a series of frustrations, disappointments; there were no emotional satisfactions with people. I lived in books, valuing only science and skepticism. My heroes were Freud, Einstein, Darwin—and the world's great atheists, such as Marx. Never can I recall having believed in God, yet the need to do so must have been in me, as Jung speculated: a basic need to feel a part of a larger whole, that the universe is ordered. I had believed the universe was chaotic, and the order that we perceive was man-made because of the requirements of human perception for meaning and closure. Yet, I now know, I created a God. When I began psychotherapy I did so with a "will to believe" in a magical-healer-therapist-God. And believe him I did! I saw him as infallible, and I literally believed his most outlandish statements. I saw him as benevolent, and I ignored the most obvious evidence of human self-seeking and pettiness. I saw him as omnipotent, and I was blind to his dependence upon people who were dependent upon him. It was a horrible cosmic joke that out of unconscious needs for a God I selected a therapist who had unconscious needs to be worshiped, and who was compelled to surround himself with worshipers and followers. Yet it was not bad luck alone, nor was it the synchrony of unconscious processes, though both were involved. Rather, it is a pervasive human problem to realize our own frailty, helplessness and cosmic insignificance in an indifferent and potentially hostile universe, then to repress such awesome knowledge and long for a benevolent God "out there." The religious conversion comes next. It is a process which makes what Eric Hoffer calls "true believers," whether the commitment be to Christianity, Judaism or, as in my case in childhood, to a devout atheism. Though our individual experience may not accurately portray the other person, and although our thoughts about ourselves may also be delusional, our own experience is all we have. God, as the Buddhists have maintained for centuries, is within us. Put another way, there is no power for anyone which is not ultimately illusory except that power he can find in himself and, paradoxically, that power may come only through a recognition and acceptance of one's own powerlessness.

CHAPTER TEN

Lucy As Co-Therapist

Re-entry

W HEN I returned home from Esalen I had reality problems. I was a different person but the world was the same. I had to extricate myself from the complicated legal, administrative, and professional ties to my therapist. This I did.

It was very satisfying to me that when I returned Lucy was the same. She greeted me with her usual hoots of joy, embracing and kissing me. My love for her had not changed, though I noted a change in my intellectual and scientific thinking about her. I was more sensitive to her moods and emotions, more empathic, able to "feel with" her. I realized that one effect of my Esalen experience was that I had "lost my mind and come to my senses"—to use Fritz Perls' phrase. That is, I could still cerebrate at will, but my intellectuality did not impede sensory or emotional experience. I was more flexible intellectually; for example, I no longer was afraid of the "danger" of anthropomorphising in studying Lucy. To make statements about Lucy's thought processes, moods, and emotions seemed little different from making inferences about the inner experience of another human being. It is true that another human being can use words to communicate experience and Lucy cannot, unless her ASL signs are considered words. But the words are not the experience itself and, indeed, words often reflect experience inaccurately even in the most careful introspectionist. I felt scientifically comfortable talking about Lucy's inner life as long as I could clearly state (1) my inferences (2) the behavior from which they were inferred (3) the situation in which the observation-inference sequence occurred and (4) the frequency of the behavior on which the conclusion was based. Though it horrifies the strict behaviorist, I feel no anxiety about saying "Lucy likes Joe," for example, when she regularly kisses him, invites him to tickle her, and offers him food; or that she hates Don if she attacks him whenever she can. As with inferences about human experience, inaccuracies occur because behavior may conceal motives as well as reveal them. I think Lucy has upon occasion concealed her dislike for someone by a kiss and bitten them later, in a more appropriate situation. For the most part, however, Lucy's sounds, gestures, movements, and actions in specific situations seem highly correlated, much more so than in most human beings. Therefore, it is easy for Jane, Steve, or me to read her moods. People who do not know Lucy will often misread her emotions, even when they seem obvious to us. For example, when Lucy approaches someone affectionately she may jump into their arms and kiss them. As her face approaches within two or three inches of the face being kissed, some people are frightened by her teeth, feel they are being attacked, and pull back. The pulling back may change Lucy's feeling from affection to anger. Generally, her emotions are clear, consistent, highly developed, and "humanized" by comparison with chimpanzees in nature.

In her book *In the Shadow of Man,* Jane Goodall said that chimpanzees show a lack of consideration for one another's feelings. She said she could not conceive of chimpanzees developing emotions for one another comparable to ". . . the tenderness, protectiveness, tolerance, and spiritual exhilaration that are associated with human love in its truest form." Dr. Goodall was talking about chimpanzees in nature, and probably about the relationship between adult male and female, whose copulation never lasts longer than fifteen to twenty seconds once intromission has occurred, and where there is no durable bond between male and female analogous to human monogamy. In any event, that description does not fit Lucy. Her behavior in certain situations is sufficiently consistent to permit the inference that she experiences definite emotional states.

The emotions she exhibits most clearly are affection, anger, fear, joy, tenderness, greed, jealousy, anxiety, concern, protectiveness, and many others. The affection she exhibits towards Jane, Steve, and me is sufficiently intense and enduring that I would not hesitate to call it love. It is an affection that is always there, side-by-side with a protectiveness and concern for us that is touching and tender to see. Since Freud we have known that human love is ambivalent, co-existing with the potential for hatred. Love between parent and child is especially ambivalent as the structure of the relationship means that one is often frustrated or hurt by the actions of the other. I have never seen this kind of ambivalence in Lucy toward Jane, and Lucy has been hostile toward me only two or three times, a record few father-daughter relationships can claim.

Her anger and her fear are obvious when she experiences them. Anger makes her erect her hair, so that she seems much larger, and her back is unmistakably savage as she sways from side to side, preparing to flee. When she is frightened she runs, screaming, and occasionally her behavior becomes so disorganized by fear her emotion could be called terror, as when she screamed, grabbed the steering wheel with one hand and threw the other around my neck as we were driving over a high, narrow bridge.

Her tenderness, protectiveness, and concern toward her cat and Nanuq were both obvious and moving to those who saw them.

Furthermore, Lucy had developed an awareness of our emotions. I am well aware that I could not prove these statements by the epistemological criteria of experimental psychology. Nonetheless, I can relate my experience that Lucy seems very sensitive to our emotional states, and that she modifies her behavior in terms of what she "thinks" we are feeling. This is particularly true of her relationships with Jane, which is characterized by protectiveness, tenderness, and concern.

If Jane is distressed, Lucy notices it immediately, and attempts to comfort her by putting her arm about her, grooming her, or kissing her. If I am the cause of the distress, for example, if we are arguing, Lucy will attempt to pull us apart or to distract me so that Jane's distress is alleviated. If Jane is sick, Lucy notices it immediately. For example, on every occasion when Jane was ill and vomited Lucy became very disturbed, running into the bathroom, standing by Jane, comforting her by kissing her and putting her arm around her as she vomited. When Jane was sick in bed Lucy would exhibit tender protectiveness toward her, bringing her food, sharing her own food, or sitting on the edge of the bed attempting to comfort by stroking and grooming her.

Sue Savage regularly took Lucy outside to play, or for rides in her sports car, and one warm day in July she wrote:

Lucy had her kitten back today and was unbelievably calm. We went outside and there was no running, *not once*. I cautioned her about sneaking away like yesterday and there was no hiding. She took her kitten around bushes and trees, moving very slowly and cautiously—supporting the kitten at all times very carefully. She was so cautious she even looked at me for approval after putting her hand on an apple in the apple tree. She did not pull it off till I nodded my head. Her movements and care of the kitten were just beautiful, absolutely beautiful, today. It was one of those times where you're really feeling the essence, the soul of the chimpanzee. My respect for Lucy grows immeasurably on days such as these. It's always sobering, I feel, for a human being to see a chimpanzee display so deftly those behaviors such as consideration, trust, patience, understanding, etc. that we presumptuously have called "human traits."

Sue was one of the brightest of the people who studied Lucy and was generally recognized as the top graduate student in the psychology department. For several years she had a beautiful, loving relationship with Lucy. Lucy was so affectionate toward her so consistently I would not hesitate to call it love. Then, after Lucy started menstruating she became more irascible and Sue could not control her. They began to have power struggles when they went outside together and Sue decided to end the relationship for the time being. I thought it was a great tragedy, but I could understand Sue's action. She had once lost part of her finger to an adult chimpanzee in a colony and hesitated to risk additional injury.

Lucy reads my feelings quite accurately even when they are separated from my words, though she is not as protective toward me as she is with Jane. For example, I might tell her to stop doing something—to stop attempting to get me to chase her, to get off the couch or to put her coffee cup back back on the table instead of leaving it in the middle of the floor. If I am genuinely behind my words, that is, if I am feeling a determination to make her do what I have asked her to do, Lucy will obey. But if I am simply saying it without feeling genuine determination, saying it in a half-hearted manner, Lucy will not obey.

If I am ambivalent, she seems to perceive it and bases her behavior upon it. I will be reading, for example, and Lucy will invite me to play with her. She will do this by grabbing my hand, snatching the book, grooming my feet, unlacing my shoes, or putting her head between my legs so that her bottom is exposed to be tickled. If I am involved in what I am reading and do not want to play with her, all I need do is say, "Stop it!" On the other hand, if I am not involved and would rather like to stop reading and play, she will ignore my "Stop it" and continue whatever she is doing.

When I am sick, Lucy is also concerned and solicitous. For example, when I am nauseated it disturbs Lucy and she tries to comfort me. Once when I had a severe attack of vomiting Lucy forced open the bathroom door, made distress sounds, rocked, touched me reassuringly, tried to kiss me. When my severe retching continued and none of her actions had reduced my distress she attacked the toilet—screaming in rage, slamming the toilet seat and lid down, and striking a hard blow with her hand against the lid. This was her typical response to either Jane's or my vomiting, but one time she behaved quite differently.

On that occasion I had had too much to drink, and the nausea was alcohol-inspired. When I started vomiting Lucy rushed in distressed and concerned as usual, but she seemed perplexed. I vomited briefly then went to the bedroom to lie down. Lucy followed me, watched me undress and get in bed, touched me gently as though to reassure herself that I was all right, then made a beeline for the bathroom. She stood by the toilet and tried to make herself vomit. She opened her mouth wide, stood on two legs, leaned over the toilet and made gagging sounds as though trying to imitate my vomiting. I do not understand this behavior but since she clearly was distressed by my vomiting before she imitated it, I speculated that identification can be a defense for chimpanzee children as well as for human beings.

Lucy's sensitivity to emotions is one reason I decided to try an experiment—to bring her into an unstructured group and see what would happen. There were other reasons too, arising out of my trip to Esalen.

Lucy as Co-Therapist

After that first Esalen trip I was impressed with the power of the group to facilitate change without as much risk of transference counter-transference distortion as I had experienced in individual therapy. I was particularly interested in the power of nonverbal exercises and experiences, if these could be integrated with the psychoanalytically oriented psychotherapy that I practiced. To facilitate character-change and not symptom-change alone the powerful cathartic and abreactive experiences produced by the encounter group would have to be integrated with cognitive processes. Intellectual functioning then would not act as a defense against emotional experience, and powerful emotional experience would join with, rather than overwhelm, the intellect. Reflecting on my Esalen experience I decided that one precipitating factor was the complete, "far-out" lack of organized and conventional structure. With Lucy in the group there would be no organized and conventional structure to limit tolerance for new experience. So Lucy was going to be a "co-therapist" on an experimental basis.

The groups I was conducting at that time met in my home. I talked it over with the group first, of course. I told them that I had never done such a thing before, that it was an experiment which could be very powerful since Lucy, having no words, would have to relate to them by responding to their emotions and body language, and that the formalized perceptual rules they followed—their belief systems—would not be binding on her. She could do nothing else, since she had no words herself even though she could think; and that she might therefore be reacting to unconscious parts of themselves and that this could be very disturbing. I also told the group that it could be dangerous, and that I could not guarantee that someone would not get bitten though I did not think that this was likely. I told them also that here—as at any growing edge of differentness—there was a potential for great joy, too; or for a "far-out" and different experience. I thought a species-isolated and humanized chimpanzee had never been a participant (or a leader) of an encounter group before, and I very much wanted to see what happened. Each member of the group agreed.

At that time Lucy was three years old and weighed thirty-five pounds—thirty-five pounds of dynamic energy and the muscular strength of the average hundred and seventy-five pound man. We were sitting on the living room floor in a circle when Lucy rushed in, a flash of rapidly moving black skin and hair. She ran around the outside walls of the room first, then approached the group, going from person to person. She stopped momentarily at each one, smelled them for a second or so then moved to the next one. After she had briefly smelled and looked into the face of each person and had achieved some familiarity with the group as a whole, she began to interact with specific group members on a more individualized basis.

She stopped first before Betty, a woman in her middle twenties. Betty was beautiful and extremely seductive; yet I had always felt she was enormously hostile beneath her highly sexualized way of relating to everyone and everything. Betty had enormous breasts and had been sufficiently preoccupied with them to have tried to calculate their weight with a bathroom scale placed on a chest-high table. She had confided to the group they weighed "over five pounds each." In any culture but Mother's Day America she would have been considered less attractive if not deformed. Lucy stood before her, looked into her eyes, then tapped her lightly on the breasts. Betty giggled, but said nothing and did not stop her. Lucy paused for a few moments as though considering her next action, then continued to poke at Betty's chest. Betty's giggling continued though it seemed a bit more forced. Lucy continued hitting her breasts a little harder each time and I began to get worried. Then Betty tried to embrace and kiss Lucy. Lucy quickly wiggled away, would not allow Betty to embrace her, and continued to hit her on the breasts. At that point I interrupted. I told Lucy to stop it. I could not tell whether Betty looked disappointed or relieved, but I felt that her inability to respond to Lucy's aggression with anything but seduction would get her bitten. (When we discussed this later Betty understood that the vignette with Lucy was the story of her life; she was terrified of aggression and hostility in any form, was unaware of any in herself, and responded to all external aggression with seduction.)

Lucy's next client was David, a young obsessional man; a graduate student in psychology. David froze. His face became a mask of white immobility as Lucy began to push him in much the same manner as she had pushed Betty. By now she felt more at home in the group and was laughing as she hit his chest, arms, and legs. The more Lucy hit him the lower his voice became, until he was telling her to stop in a voice so low it was all but inaudible. Lucy pushed him on the chest, threw her arms around his neck, climbed on his shoulders, stood on his head then jumped to the floor. She literally walked all over him—as did every human female with whom he had been involved. There was a reciprocal relationship between Lucy's aggression and David's passivity. The more aggressive Lucy became the more passive David became; conversely, Lucy's aggression increased in direct proportion to David's passivity. I stopped this interaction with Lucy for I was afraid he would be bitten. Lucy was so clearly dominating him, though she was no more than one-fifth his size, that I felt his passivity could only be hiding an enormous hostility, and that with no place to go outside David it could lead only to depression, which is exactly what happened. Shortly afterwards he became quite depressed—as he did after, as he put it, "any kind of hassle with my wife." I wonder if Lucy had not demonstrated a process that frequently occurs in sadomasochistic marriages. That is, the passivity of one partner produces aggression in the other, and both come to feed on one another's response. For example, the aggression of one partner helping him avoid tenderness while the passivity of the other enables him to avoid his own aggression as long as he can see it in the other.

David's and Betty's words had lacked the emotional force and authenticity to make an impression on Lucy. Their words, feelings, actions, tone of voice, etc. had not been correlated. The discrepancy upset Lucy who was visibly becoming more anxious, and then responding to her own anxiety by becoming more aggressive. She was searching for external limits and control. I felt it necessary to provide them, which I did simply by telling her, "Goddammit, slow down and sit still!" Since I genuinely felt what I was saying, Lucy relaxed and obeyed. She sat still while looking around the group as we talked about the experience. Though quieted to this extent Lucy still found it impossible to allow any group member to touch another. If one person put his hand on the shoulder or leg of another, Lucy quietly picked it up and placed in the lap of the owner. She did this twice. This was not surprising since Lucy has always done the same thing with Jane and me. It is as though she must keep interpersonal contact to a level which minimizes aggression, which is certainly fortunate since chimpanzees are such powerful animals. How unfortunate that we humans have no such built-in controls!

To Joe, Bill, and Susan she was very affectionate and playful. These three people were more able to be emotionally responsive, both with Lucy and in general; and Lucy responded warmly to their emotional honesty. Being more open to their own experience, Joe, Bill and Susan did not have the chronic muscular tensions that go with repressed anger, as did Betty and David. Lucy experienced their being more relaxed, read their muscles, and was affectionate toward them. She jumped from one to the other, touching their lips lightly in a chimpanzee kiss. (At that time Lucy's kisses were brief pecks on the lips or cheeks. The sustained mouth-covering kisses did not appear until about sexual maturity.) She liked them and enjoyed sitting in their respective laps. She would turn around, present her back to one of them, and "beg" to be stroked and tickled. These three people were able to play with her so that there was a friendly give and take of affection. The affection was on a nonverbal muscle-to-muscle level, and Lucy understood the taking turns aspect of it. She would present her rear or her back to be tickled, then turn around and tickle or cuddle Susan, Joe, or Bill. After ten or fifteen minutes, most of which was a repetition of what I have already described, I took Lucy back to her room, then returned to the group and we discussed the experience.

Naturally, the contact with Lucy had different meaning for each of the participants, but to me it dramatized again the necessity for learning how to be aware of and accept and integrate aggression and hostility. The people who had been unable to control Lucy even to prevent her from walking all over them, viewed their own hostility and aggression as sources of threat. They did not see these feelings as a normal and natural part of their own being, to be actualized in an aggression producing situation. They had not learned, as I had learned at Esalen, that such feelings, if accepted and integrated, can be the source of great strength and creativity. Lucy, not having been raised in the Judaic-Christian tradition, did not know that the meek would inherit the earth. She thus exhibited a spontaneous joyful reaction to people which was inherently energetic and aggressive, and which became hostile only when she was threatened by not having external limits, or when she was being stimulated by the muscle tensions and inappropriate emotions of people who had large quantities of repressed hostility. One difference between Lucy and us was not that we had "human" emotions and she did not, or that she had "animalistic" emotions and we did not. Rather, the difference seemed to be in the capacity to internalize and integrate a system of controls over feelings and emotions. The group members, or we humans in general, can internalize and sometimes integrate with the rest of ourselves entire belief systems of controls over emotions, particularly over the sexual and aggressive ones, tabooed by society. Often the process is overdone, and we become too civilized. I once knew a person, for example, who had been raised on such a diet of "If you can't say something nice, don't say it," that he could say nothing which might offend anyone, regardless of the circumstances. Each evening for twenty years he drank a martini before his meal because his wife served it to him. For fear of offending her he could not tell her that he did not like alcohol.

CHAPTER ELEVEN

Aggression and Hostility

AGGRESSION AND hostility are basic dimensions of human existence and cannot be avoided. Each individual will at some time experience these emotions in himself, and upon other occasions be the recipient of the aggression and hostility of others. Thus each person must sooner or later come to terms with this part of the human experience, and it is not easy to do. It is difficult because American culture does not clearly distinguish between aggression and hostility. They are by no means the same emotion, yet our culture simultaneously stimulates their production while sanctimoniously judging them as evil. For example, as children we are taught: "Be individuals, stand up for your rights, be aggressive, get to the top, become a success!" While at the same moment we are told "Turn the other cheek and follow the Golden Rule."

Through the internalization of such contradictory values as a people we become capable of waging murderous war against helpless civilian populations while uttering the most pious and sanctimonious rationalizations. The explicit recommendation to "Bomb them back to the Stone Age" was criticized by many military and government officials, even as they attempted to do so. As individuals, the internalization of these contradictory values can cause many different personality problems, for example the inability to be self-assertive, to act aggressively in one's own behalf for fear of an angry interpersonal confrontation.

Lucy provides a unique opportunity to study aggression and hostility. She is a member of the animal species closest to man and yet she was raised in isolation from her own species, and her aggressive and hostile behavior can be compared to that of chimpanzees in captive colonies and in the wild.

First, it is necessary to distinguish between aggression and hostility.

Aggression is energy expressed in the direction of the environment; in behavioral terms it is activity. It is not hostility because it is not an expression of a wish to hurt or destroy. When Lucy runs through the house, rolls a basketball across the floor, bangs the wall, jumps over furniture, and the like, she is being aggressive, but she is not being hostile. Aggression can become hostile, as when actively playing children cross some physical or psychological boundary and suddenly start to threaten or fight. Before that point the activity, however aggressive, is better described as self-assertive behavior rather than hostile behavior. It is impossible to deal effectively with Lucy without making this distinction.

During the first year of her life Lucy was fairly helpless. For most of this period, she showed aggression, but no hostility. For example, she very actively explored her small world. She reached for and studied objects, put them in her mouth, examined them with her hands—demonstrating a capacity to effect slight changes in her world by activity, but without hurting or destroying. If frustrated she had temper tantrums—she screamed, choked, and sputtered, tore her own hair, and turned somersaults rapidly—still not being destructive toward another organism or the environment; thus being aggressive but not hostile. The first instance of aggression which might also be hostility occurred at 5½ months. A friend brought her two young boys for a visit to see Lucy. One of the boys was five and the other one was seven years old. When Lucy first saw them—with no provocation whatsoever that we could see—she barked, a sound we had not heard her make before. Her hair stood erect, and she swayed repetitiously from side to side in what seemed to be the rudiments of the threatening display shown by older chimps. This first example showed the difficulty of the distinction between aggression and hostility. Lucy may have had no wish to hurt or destroy those two little boys, even though she uttered the same bark she utters when obviously hostile. She could have been using aggressive behavior as a threat to maintain peace and tranquility, barking the equivalent of "Stand at attention! Notice me! You are now in my room!"

Social Status

As Lucy grew older, it became clear that she could categorize, could group together, and that her aggression and hostility was channeled by her categorization of people. The most important categorization was family membership—we or they—and next important was rank.

There has always been more hostility toward outsiders and strangers; Lucy has never bitten a family member. People outside the family automatically have a lower social status than Jane, Steve or me, and are more likely to be the objects of hostility. For example, when Steve would engage in rough and tumble horseplay with another ten or twelve-year-old boy, Lucy's response was initially aggressive toward the person Steve was playing with and then might evolve into hostility. If they were playing chase or wrestling, Lucy would sway from side to side, vocalize her threat bark, her hair would become erect, and if they did not stop the play she usually would bite her brother's playmate.

Within the family, Lucy's aggression is an attempt to stop hostility from developing. In other words she will sometimes threaten Steve or me to stop our aggression and maintain an optimum cordial level of family interaction. If Steve or I act as though we are mad at Jane, Lucy will go to Jane's side and look threateningly at us. This aggressive response in potential defense of Jane has continued throughout her life, but the discreet acts which she interprets as threatening to Jane have become increasingly subtle. She has learned to make rather precise differentiations. For example, if she is playing by herself in the living room and Jane and I are discussing some intellectual topic, she ignores us. If the discussion becomes emotional and we begin to argue, she will stop what she is doing and watch us. If we show any additional emotion she will then approach and sit by Jane and attempt to groom her. If we continue to argue beyond a critical point Lucy will then approach me and try to distract me, usually by inviting me to play. If I refuse to be distracted then she might bark and threaten me.

With a stranger the slightest movement toward Jane, even though the stranger does not touch her, may bring Lucy to Jane's side. If Jane even shakes hands with somebody else Lucy might take the other person's hand and remove it from Jane's. Touching between Jane and me is likely to be responded to in this protective fashion—unless the casual touching is correlated with emotion in which case it brings an even more dramatic protective response.

Her response to rough play between Steve and me also seems determined by status. When Steve and I wrestle she attempts to separate the two of us rather than aggress toward either of us. If we ignore her and continue she always threatens Steve because I have higher status. Steve was once a state champion in Iudo, and is stronger than I am. So Lucy is not threatening the weaker one but the one with lower status. She is not doing so hostilely, as she has never bitten him, but in terms of her own rules she is maintaining the existing peaceful order of relationships.

This construction of her own rules for conduct in her house and family is one of the most interesting aspects of her being raised as a human. I am very grateful for this theory, which I learned from Sue Savage, for it suddenly made many different observations fall into place.

Sue Savage spent hundreds of hours observing chimpanzees in a colony, and also interacting with many colony chimps on an individual basis, taking them for walks in the woods or for rides in her car. She was working on a Ph.D. in psychology, specializing in chimpanzee behavior, and she became so proficient in interpreting their behavior she could handle many of them when other people could not, by knowing when hostility was about to occur, and ways of avoiding it. She worked with Lucy for several years, once or twice a week, and she told me Lucy was very different from colony-raised chimps; much more human in many ways, one of which was that Lucy constructed rules for those around her to follow, whether they were interacting with her at the moment or not.

For example, a dominant chimp in a colony tries to control the behavior of a sub-dominant animal only when the two are interacting—at other times he does not usually try to manipulate and control what the sub-dominant chimp is doing, unless his own property or territory is involved. This is not true of Lucy, Sue pointed out; she wants things to be done in a certain way whether she is directly or immediately involved or not. Her internal rules govern the behaviors she will permit between Jane, Steve, and me, or between us and the environment. When she is in estrus, for example, and generally more irascible than usual, she will allow Jane to hit me on the shoulder with no apparent interest, but if I clench my fist and hit Jane she barks and threatens me. This prejudice irritates me and amuses Jane. One day she was mad at me and hit me hard on the shoulder. Lucy watched with interest but said or did nothing. Jane then issued an invitation which I declined. She said, "Go ahead, hit me." When I made no move she added, "You coward, you'd fuck her for science but you're too chicken to hit me," which I considered less than logical and not completely true.

Everything in its Place

Lucy's construction of rules is dramatically illustrated in her response to the moving of furniture. No one may move any furniture in her living room whether she is using it at the time or not. The moving of furniture invariably produces an aggressive attempt to stop it, or an actual attack and bite. Every non-family member who has moved furniture in her presence has been bitten even though, like Shelly, he was not a stranger.

When we move furniture Lucy is clearly in a conflict situation. She tries to stop us, even when it is Jane who is doing the moving, and she becomes disturbed and anxious if we continue, which produces in turn more aggression, but it has never produced a bite.

Lucy has displayed a rather diabolical tendency to set up situations in which aggression or hostility would be justified. This was amazing to us when we first observed it. We have observed it enough times to know that it was no accident and that Lucy is capable of foresight and of planning the expression of hostility. The first time she did this she was about four years old. She was visited by eight or ten young children to have a party. Jane knew that that would be a difficult situation to control Lucy in and watched her closely. Jane told her in advance that she had to be nice and not be aggressive toward her guests. Lucy was sitting on the floor surrounded by toys and admiring children. She watched Jane watching her and she watched me. And she looked at the children. Then she slowly pushed some of her toys toward one of the children in what the child could only interpret as an invitation to pick them up and play with them. The child then did pick up a little toy and Lucy immediately barked at the child and started to bite her. Curiously, she likes her toys and plays with them much as a human child, yet they become the object of aggression after they are handled by others. This is particularly true of dolls. Lucy has played with dolls since she was very young. She enjoys playing with them and carried them in the same way that a mother chimpanzee carries her offspring in nature. She carried them on her hip or she walks on all fours carrying the doll on her back. However, she will displace aggression from a person to a doll, or often tear it up, biting it, hitting it and throwing it, after someone else has played with it. When other children at the party played with her toys (a doll, a plastic car and a plastic purse) Lucy watched them and as soon as a child handed the toy back or put it down, Lucy attacked the toy and destroyed it. To demonstrate the point I have many times taken a doll away from Lucy, stroked it and said, "Nice dolly, nice dolly" only to have Lucy snatch it back and destroy it instantly. I wonder if this is an example of the primordial origins of the scapegoat phenomena—directing aggression toward a more helpless object or person, one less likely to retaliate, than toward its source.

Controls

As a therapist I have always thought that the internalization of functional and flexible controls over one's aggression was as important a part of personality development as sexuality. With man there always seems to be a delicate balance between internalizing enough control so that aggression can be channeled and organized and social interaction possible, or internalizing too much control, for example, as with depressed people. With such people the process is often overdone so that the person is too inhibited and cannot be hostile even in situations which demand it. For example, sober alcoholics may get drunk immediately after becoming angry with their wives; one patient made a suicide attempt upon discovering that her husband was having an affair with another woman. In such instances normal hostility had been over-controlled and redirected against the self. Lucy controls aggression and hostility, but the controls are her own. She has demonstrated little capacity to acquire middle-class values and controls over these emotions. We know that she can control them, but the standards are her own.

When we are present she may control hostility and then express it in forbidden ways when she thinks we are not looking or when we are absent. For example, when Jane or I were present she would not attack Tom, an African Grey parrot I once had. Yet when either one of us would turn our backs she could not resist attacking and tormenting him.

There is one social situation which invariably produces aggression, which may turn into outright hostility. This is when Jane and another female are in Lucy's presence. The other female always has lower status than Jane. Even though she may have a warm relationship with Lucy whenever Jane is not present, when the two of them are together, the woman of lower status becomes the object of aggression. She is very likely to get bitten if she makes the wrong move, such as playing with one of Lucy's toys, moving furniture, or arguing with Jane. On several occasions Lucy has been playing with a female friend and getting along fine only to attack her the moment Jane entered the room. In other words, motherhood status alone may be both a stimulus for aggression and determine the direction of its expression. There is a sex difference which interacts with status. Females are more likely to be bitten than males.

The most dramatic thing about Lucy's control over aggression and hostility is that she has not acquired our controls, but that she has had her own from birth.

Like all chimpanzees, Lucy had from birth a behavior designed to reduce aggression. It is called the pronated wrist and consists of extending the arm with the wrist flexed so that the hand hangs limp, a position which chimpanzees interpret as supplication, a plea to cease and desist aggressing. In a sense the hand is being held in a position in which it could not be used in attack. In nature, the gesture is usually successful if the aggressor is not simply carried away with his own aggression. If he is charging so fast and with so much force that he cannot stop himself, the smaller chimp screams and runs. Otherwise, the pronated wrist gesture is an effective aggression-reducer. Lucy uses it when she feels it would help inhibit the aggression of an adult. In situations where it is not likely to inhibit aggression she uses a learned behavior. I first discovered this in a painful but amusing way.

Our house is full of trees and plants, and Lucy knows she is not supposed to touch them. Generally she is obedient, having learned since childhood that Jane and I become angry if she destroys plants. I once had a banana tree of which I was particularly fond. I had grown the banana tree in the courtyard and it was five feet high, growing in a tub. I had hopes that eventually it would grow much higher and bloom. One evening when we were having company I moved the banana tree from the courtyard to the living room. There were many other trees and plants in the living room as well, since we were expecting dinner guests, and wanted the house to look as nice as possible. Jane and I were both anxious about this dinner party because Lucy was to attend. We wanted her to be on her best behavior.

Jane and I were having a drink in the kitchen before any of the guests arrived. We were discussing the dinner party. I finished the martini I was drinking and could feel it begin to warm me when I noticed that Lucy was missing. I went into the living room and was appalled to see Lucy sitting in the center of the floor surrounded by the remains of the banana tree. She had torn every leaf off it, had taken bites out of them, and had even dismantled the trunk of the tree and emptied the soil in which the tree had been planted onto the floor. It was a scene of chaos and guests were due any minute. I lost control, screamed "Goddamn you!" at Lucy, and raised my hand to strike her, cursing with all the invective I could command. Instead of extending her pronated wrist, Lucy looked me directly in the eye, smiled her little girl smile, and touched her nose with her thumb, forefinger extended in the ASL sign which means, "I'm Lucy." I stopped in mid-gesture! I could not hit her, my eloquent chimpanzee daughter. I think social learning had given her a more effective gesture for appeasing me than evolution. So Lucy was not punished in any way for this incident, and Jane and I got the mess cleaned up about the time the first guest arrived.

Since that incident we have observed that Lucy inhibits our aggression in many different ways, not simply by the pronated wrist. Sometimes, for instance, if I am mad at Lucy she will run toward me rather than away from me, throw her arms about me and kiss me before I can render harsh words. Most frequently, though, she attempts to inhibit my aggression or Jane's by distracting us—by acting cute, making faces, turning somersaults or twirling around and around until she collapses from dizziness. She has an uncanny knack for reading anger and is highly skilled at defusing it.

To illustrate Lucy's sensitivity to the repressed or potential anger in other people, there was the case of Jennifer. Jennifer was a woman of thirty who had three children. She was in a group which met at my house, and she had social contact with Lucy from time to time. Lucy did not like her, and even though Jennifer wanted more contact with Lucy I kept them apart expecting Jennifer to get bitten. This was Jennifer's behavior in the group.

Jennifer felt that her husband neglected her and that her three children "did not appreciate me." She had an unlined, vacuous smiling-through-pained facial expression. As she put it, she had been a "victim of spastic colitis for many years." No physical cause had ever been found for her spastic colitis nor had a cure been achieved. She had had two bouts of exploratory surgery before a physician told her it might be psychogenic in origin and referred her for therapy. She described herself as happily married, even though she "regretted" her husband's regular absence. In the group she was a reassurer of other people. She tried to comfort them whenever they cried or got angry, often in ways which reduced or impeded the other person's capacity to experience the free flow of his own tears or anger. She was totally paralyzed the day that Lucy was in the group. In a later session—one in which Lucy was not present—she suddenly said that she had a pain in her gut. I asked her to be the pain in her gut, to lend the pain words, and to let her consciousness flow toward the pain rather than away from it. As she "became" the pain in her gut "out loud" she at first had difficulty and instead talked about it. She said, "I have the feeling that my guts are twisting." I then gave her a towel, and asked her to twist the towel. Her face reddened as she did so. I then asked her to make sounds as she continued to twist the towel. The sounds she made were angry and I asked her to continue to twist the towel while loudly saying, "I resent my husband." That sentence was the trigger. The dam broke and hostility burst forth. She screamed, cursed, kicked at her husband, cursed the physicians who had operated unnecessarily, pounded a pillow and tore the towel, her hostility exploding out of control. From her roof-top room Lucy picked up the energy and pounded and kicked the walls herself, uttering loud regular whoops. Jennifer continued until exhausted. She then looked surprised and somewhat pleased and said that she had never realized that she had felt that way. While this was dramatic to see I did not consider it more than a step on the road toward wholeness for Jennifer. It was a cathartic experience and produced dramatic symptom relief. Her colitis stopped immediately, but as a person her growth had to continue before hostility would become a permissible experience for her, a part of the self which could be accepted and integrated without translating into physical symptoms. In any event, after this experience whe was able to relate to Lucy without paralysis. She lost her fear of being bitten and Lucy liked her better.

We have never observed any *chronic* redirection of Lucy's hostility from the environment back against herself, as was typical of Jennifer. The word chronic is necessary because occasionally Lucy will become so angry under circumstances where the anger cannot be directed that she has a temper tantrum. She will then scream, pull her hair, or scratch herself. Other than this, however, we have never seen her direct aggression against herself. In human beings, suppressed aggression often is redirected against the self and produces characteristic muscular sets. Yet we have seen no evidence of a character armor in Wilhelm Reich's sense. Lucy's posture and her movements are the same as chimpanzees in nature when compared with the movies taken by Jane Goodall of free-living chimps in the Gombe Preserve of Tanzania.

"Rolfing" is a system of deep muscle massage named after its developer, Dr. Ida Rolf.

One time at Esalen while I was being Rolfed and experiencing the pain of chronic contraction in striated muscles being relieved by deep massage, I discussed the muscles of chimpanzees with my Rolfer, Seymour Carter. We reasoned that chimpanzees would have none of the conflicts which, in man, may get concretized as inflexible muscle sets or tensions; for example, buried rage as a chronic contraction of the shoulder muscles, or inhibited sexuality reflected in the rigidity of the pelvic muscles. We predicted that since Lucy had less need and capacity than man to internalize psychological conflicts that she would have no need for Rolfing, that she would have no "charley horses," or hard knots in the midst of otherwise flexible muscle tissue. When I returned from Esalen I checked out this hypothesis. Lucy's muscles were firm and at the same time flexible. There were no charley horses or hard spots. When Lucy relaxed all of her muscles, all of the parts of each muscle felt relaxed to the touch. There was no evidence that "knots," muscle spasms, or adhesions between the sheath and the muscle itself had formed, that is, nothing which might indicate chronic psychological conflicts concretized as a muscular armor. These observations were later confirmed by a Rolfer, Stacey Mills, who visited and felt Lucy's muscles. I think Lucy's muscles have this quality in part because she has no need to inhibit self-assertive behavior and less need than we do to inhibit aggression.

Fear of Aggression

In Freud's day, neurotic problems centered about sex—particularly sex related to growth within the family such as Oedipal problems. Today, however, while such problems certainly occur they are less common in the practice of psychotherapy than people who could be described clinically as aggression-blocked. That is, they are unable to tolerate the overt emtional experience of anger, hostility, or hatred, or to assert themselves aggressively in their own behalf. It was incredible to see how few visitors to our house could mobilize sufficient aggression to protect themselves from Lucy or to control her, even when she was very small.

Lucy always had a deep fondness for hats, cigarette lighters, pipes, and women's purses. Whenever a visitor would display one of these articles she would be fascinated with it and try to get it. Very few people could say "no" and make it stick, even when Lucy was only two or three years old. She would climb all over a person, sit in his or her lap, and try to get the pipe, glasses, cigarette lighter, or whatever goodies purse or pocket might contain. If denied her prize she would push the person and very few people would push back or set limits. Perhaps I dealt with a biased sample, but I remember a six-foot youth of twenty-five, a Vietnam veteran, who was scared to death of Lucy. She once took his hat off his head after he told her not to do it, held it to her rear, defecated in it, and handed it back to him as though she were giving him a gift. He did not get angry or show any other emotion. His face whitened, but he said nothing to Lucy. Both of us were embarrassed and I promised to buy him another hat. This kind of inability to respond appropriately to open aggression on Lucy's part was particularly characteristic of "nice people," often the children of hyperintellectualistic or religious parents. They would say, "Don't do that, please; it's not right, okay?" Or instead of saying, "Get the hell off my lap!" they would say, "That is a hat, it is for wearing on my head, not on your head." I think such people suffer from a fear of their own aggression or they could have blown up at Lucy. Their passivity in the face of Lucy's aggression toward them was an attempt to be "nice people" who do not have aggressive or violent emotions. They are the ones most likely to be bitten.

Occasionally Lucy seems afraid of her own destructive potential. I cannot prove this, but her behavior in such cases is consistent. First, it is easy to tell when Lucy is frightened. She will scream or rock or clutch anxiously at Jane or me, and I have seen her scream in fear when her aggression has been simultaneously mobilized and inhibited. For example, one day we were driving through the countryside on a dirt road, enjoying the beauty of the trees as they turned golden and red in the early fall. We were driving only about ten miles an hour when we noticed two children approximately Lucy's own size playing at the side of the road. While Lucy is kind toward smaller children she is often aggressive toward those her own size or larger. Lucy was sitting between Jane and me in the front seat and the window on my side, where the children were, was rolled down. Jane was driving. Lucy saw the children and immediately jumped toward the open window with a threatening gesture—her hair was erect, her teeth were bared, and she was vocalizing her threat bark. Jane said simply one word, without raising her voice. That one word was "No!" Lucy screamed with terror, then began to whimper and rock. The threatening behavior disappeared the instant Jane said "No," and I had the impression that she was screaming with the fear that she might lose control and not be able to obey her mother. This same behavior had occurred when Lucy was play fighting with Steve or me. There would be a time when the play would become so violent that Lucy would scream, and I had the impression that if Steve or I did not immediately stop, she would have lost control and attacked us—something which she certainly does not want to do.

Lucy loves rough and active games and she easily distinguishes between an aggressive game and hostility. This distinction always involves categorizing family members as "we" and friends or strangers as "them." For instance, her favorite games are aggressive—wrestling, chase, play biting, play fighting. While she will play these games both with family members and non-family members, play is more likely to turn into aggression with non-family members. Any one of us can start an aggressive game with an action which would result in an immediate attack upon anyone else. Lucy can be sitting reading a book, for instance, and I can walk up to her and hit her. She immediately starts to laugh, grabs me, and a game of wrestling has begun. She can distinguish between an aggressive game and real aggression even when weapons are used.

A bataca is a club about two-and-a-half feet long and ten inches in diameter which has a handle and a sheath protecting the hand which holds it. It is made of some firm material so that it holds its shape, yet the material is soft enough that when one is hit it does not hurt, even though the blow is quite solid. I have two of them around the house and sometimes use them in groups. On many occasions I have walked up to Lucy, raised the club high, and hit her solidly on her head or back without warning or provocation. This would result in an attack if done by a non-family member. When I do it, play results. She tries to take the bataca away from me and hit me with it, or she grabs my legs in a tackle, or she simply signals in her sign language, "Chase Lucy!"

These aggressive games constitute another indication that Lucy can control her aggression. When we wrestle I like to hold her and bite her ears, neck, stomach, or whatever I can reach. She enjoys this thoroughly and is laughing most of the time, but whenever she wants to stop the game, a different expression crosses her face and she is instantly free. She either lifts me up and throws me back or squirms so violently that I cannot hold her. This sudden show of strength led to the invention of another game we play called "Squeeze." I will simply take one of her hands in mine and slowly squeeze it. I will gradually increase the pressure until I am squeezing with all my might. Lucy will then sense when I have no more strength and begin to laugh, or then sometimes she will squeeze me back so hard that I ask her to stop.

I also taught Lucy the encounter group technique of arm wrestling. We will each put our elbow on the table, our forearms more or less in line, and grasp each other's hand. I then try to push her hand one way and she to push mine another. There is no contest whatsoever. She laughs, goes along with me for a few moments, and then easily pushes my hand down.

"The Chimpanzee is the Missing Link
Between Man and the Human Being"
—*Konrad Lorenz*

Lucy has always directed her hostility outside the family, and she has bitten people more often in the presence of family members. These observations are consistent with the theory of Konrad Lorenz that, unlike man, chimpanzees and many other animals have biologically built-in resistances which limit the expression of hostility against their own species. While occasionally a chimpanzee in nature may attack another, quarrels are usually ended by threats or social posturing rather than actual combat. Unfortunately, there are no such aggression-inhibiting devices in the glands or genes of mankind.

Once when Lucy was about nine she almost bit me. She was in full bloom at the time, her genitals a swollen pink mass of rounded bottom, and Lucy was maximally irascible. She was clearly demanding attention, and Jane and I were ignoring her. We were standing in the kitchen, our arms about one another in an affectionate embrace which excluded Lucy. Several times Lucy tried to separate us. She put her hand on my shoulder. I brushed it aside. She whimpered at Jane, who also ignored her. Then Lucy did one of the things which always gets my frantic attention. She jumped from the range, on which she had been standing, to the telephone desk. She picked up the receiver with one hand, held it to her ear, and with the forefinger of her other hand she started dialing. Afraid she would accidentally dial Tokyo or Tel Aviv, irritated by her constant interruptions, and feeling chimps should be seen and not heard, I snatched the phone from her hand and cursed her. She turned away for a moment, then turned around and charged angrily, screaming. In an instant she had my forearm in her mouth and was biting down. But the scream of anger quickly changed to a scream of terror, and before she had broken the skin she had released my arm and was withdrawing, whimpering. I believe she was not afraid of me—I certainly was not going to retaliate—but that she was terrified of her own impulse to bite me, for there was no force external to Lucy which could have stopped her. Analogous to her taboo against incest, she seems to have a taboo against attacking family members, for which I am grateful.

"Power Corrupts and Absolute Power Corrupts Absolutely"

—Lord Acton

The human being does not handle power well; with it he becomes dictatorial or megalomaniacal, or both. Since Plato first conceived of one I doubt that a benevolent despotism has ever existed, or if it did that it did not survive for long. Dictatorships are far more frequent in the history of mankind than democracies. Even within democracies such as ours, based upon a separation of powers, there have been enormous abuses. Whenever one human being has had power over another or over a class of people, he for the most part has not used it benevolently. Whether the subjugated were Jews in Germany, Blacks or women in America, Catholics in Ireland, or whatever—their rights and privileges were reduced in direct proportion to the amount of power possessed by the dominant person or class.

Since Lucy behaves in much the same way it produces the speculation that there may be a biological basis for the inability of the chimpanzee and man to handle power in a way which preserves the dignity and well-being of those who have it and those who do not.

Each time Lucy has encountered another animal smaller than herself for the first time she has threatened it or attacked it. Her first response to cats, to Nanuq, to two border collie puppies I had, and to rabbits, squirrels and strange dogs we encountered in the woods was always the same: a threat bark, piloerection, bared teeth and in several cases, chasing the small animal as well. When she encountered animals larger than herself, cows and horses, for example, her response was also aggressive from a distance, or even up close if the animal's back was turned. Then she would throw a stick or rock at them, and turn and run. If the animals were locked in a corral she could perceive their helpless state and would devil and dominate them, as mentioned earlier, until Jane or I stopped her.

With her cat and with Nanuq, we were able to inhibit her initial aggressive response and eventually a relationship of loving friendship was established. Even so, the potential for dictatorial exploitation always was there. If she wanted to go upstairs and wanted her kitten with her, it did not matter that the cat wanted to eat. She simply grabbed it and took it with her. Though she and Nanuq are friendly playmates, if the impulse appears, or if there is a sudden disagreement, she will quickly strike Nanuq a hard blow, then run from Nanuq's attempt to retaliate. In short, she places her needs first—regardless.

I am reminded of Freud's goal, so rarely achieved, "Where id was, there shall ego be." With Lucy (if not for most people as well) there is always a delicate balance between ego and id, and impulsive behavior may destroy reasoned policy at any moment, moral values to the contrary notwithstanding.

Lucy has also had power over parrots and minnows and the results were much the same. She takes a thoughtless delight in playing with a helpless, interesting organism. Her motives may be curiosity rather than sadism, as with small boys learning about death by tearing the wings off butterflies, but the effect on the sub-dominant organism is the same.

I am fond of fishing off my front porch and I often use minnows for bait. It is difficult to keep Lucy with me even though I enjoy her company because she is so fascinated with minnows it is hard to keep her from reaching into the minnow bucket. She will hold one in her hand and poke at it, unable to deflect her attention from the slippery squirmy fish in the palm of her hand. Or, she will put it in a bucket of water, watch it swim around, dip her hand in the water and chase it in ever faster swirls, around and around, until she has either crushed the minnow or splashed all of the water out of the bucket. During these activities her face assumes an expression that I would describe as devilish glee. Her eyes glisten and she is grinning and looking out of the corners rather than directly out of her eyes.

On many occasions I have seen Lucy attempt to dominate and tease parrots. I had to keep these birds carefully separated from Lucy as she is unable to resist the urge to dominate them. She pokes at them, grins at them, tries to catch them, jumps for them, and attempts to pull out their feathers.

My two favorite birds are a pair of Banksian cockatoos. These are black cockatoos with red tails. They are very rare in the United States though they are still fairly plentiful in Australia. These two were bred in captivity by a Mrs. F. H. Rudkin of Fillmore, California. I traveled to California to bring them home with me as babies beneath my seat in a commercial airplane. They were so prized that I built a cage twelve feet long, six feet wide, and eight feet high in the center of our courtyard so that the birds would never be exposed to the risk of being caged outside the house. I was horrified one morning as I was taking a shower when I heard them screaming. I rushed out nude, ran through the house into the courtyard, as I knew in my heart what had happened. Lucy was inside their cage trying to catch them. The birds were flying back and forth as Lucy, both hands flaying about, grabbed at them. So far they had managed to avoid her grasping hands, but while I watched she grabbed one by the tail while it was in flight. The beautiful red tail feathers promptly came out. I rushed into the cage and stopped what would have been a massacre. I took Lucy out with a feeling of great resignation. It was hopeless to punish her. She simply could not resist deviling them, and punishing her would have no effect whatsoever, or at least no positive effect. It would have the same effect that it does on criminals, that is, make them more bitter and hostile.

One day Lucy found a key and entered the bird room and did the same thing in a flight with three Leadbeatears cockatoos. She has been able to dominate and scare half to death every parrot I ever had except one, named Tom Bombadill.

Tom was an African Gray parrot, generally recognized to be the most intelligent and the best talker of all the parrots. I do not know Tom's early history, but whatever it was he had developed far more ego than any parrot I have ever had. He loved me but hated women, and bit them on numerous occasions. He would even fly off his perch to attack women, but he was extremely affectionate toward me. His attitude in many regards was so consistent that it seems reasonable to infer the existence of that hypothetical construct we call ego. For example, Tom was very resistant to being shown off. He was a good talker, and I taught him many charming phrases. Tom could say, "Birds can't talk," in my own voice. He could whisper, "I love you," and "I am the smartest bird in the world." I taught him these phrases with very little effort. I simply repeated them a few times as I was holding him on my hands, or standing in front of his cage. Reinforcement seemed to have very little to do with it. I was proud of his linguistic ability, and when guests would come I would want to demonstrate his speech. I would stand in front of his cage, talk to him, urge him to talk, and repeat the phrases myself. Tom would stare at me, look at the guest, stare back at me, and say nothing. To my knowledge he never talked when I was attempting to show him off. Then, when no one else was around, he would talk fluently, uttering these phrases with great frequency and enviable articulation. Friends could hear him talk but to do so they have to fool him by remaining out of sight. If he knows that we are listening he does not talk.

Tom lived in a small cage rather than in a flight, either in the living room or in the courtyard, and Lucy was always attempting to dominate him, though she was not able to do so. She would stand in front of his cage bipedally. Her head then would be at his height, and she would poke a finger at him through the wire mesh of his cage. Tom would not back down an inch. He would simply open his mouth and bite the finger. Lucy would then withdraw it before Tom could bite her again. One day Lucy picked up a broom and started beating on Tom's cage, which eventually made Tom flutter so much I was afraid that he would hurt himself. After that I moved Tom into the courtyard and kept him and Lucy apart.

My career as a parrot fancier and amateur breeder came to a dramatic end, but I learned much in the process. For fifteen years I had tried to breed the rarer species of parrots by using well-established scientific principles. That is, knowing that in nature the breeding of such birds is influenced by temperature, rainfall, humidity, light, and diet, I manipulated these variables more or less systematically with very limited success. I bred only those species which breed most easily in captivity, none of the rarer birds. Then a friend, Mr. Bob Berry, curator of birds at the Houston Zoo made a suggestion which was to produce almost immediate breeding success. He said, "Try to read the bird's mind!" Paradoxically, though I had been a therapist for many years I had never thought of taking each parrots individuality into account in this way. Bob elaborated on what he meant. Each individual parrot has a personality of its own with consistent likes, dislikes, and biases, since his behavior is not exclusively determined by his membership in a given species but is influenced by the individual bird's personal experience.

I immediately started putting his suggestion into effect. I watched each pair of birds and tried to decide such factors as these, "Do they like one another? Do they seem happy? Where in the flight do they spend their time? Do they seem to like their nesting box, and so on." If the answer to any one of these questions was in the negative I then moved something. I either changed partners, the type of next box, or its position in the flight, and kept changing something till they seemed happy. Almost immediately the birds began to breed. Within four to six months one pair of Moluccan cockatoos had produced three young, and I had six baby Lories; the other cockatoos were also nesting. The Lories and two of the Moluccan cokatoos made it to maturity, though I lost one of them. All the birds seemed happy and every bird on the place was breeding when I developed an acute infection. Somehow I had an allergic reaction to the antibiotic, after which my body chemistry was changed—at least I assume that was the cause—for I could no longer go near a parrot without getting sick. I had developed an allergy to the powder which sloughs off from the feathers of parrots. The allergy was so severe that I could not go in my bird room without becoming acutely nauseated, often vomiting and developing cold chills. So I had to sell the entire collection. Many of the birds had been friends for years and it was difficult to part with them. There were compensations, however, because I felt I had discovered something about the breeding of rare species, and had indeed demonstrated one point. It had been widely believed that birds imprinted on human beings, that is, raised by humans from the hatching of the egg to maturity, did not breed. One of my successful breedings of Moluccan cockatoos was with a tame, imprinted female, who incubated her eggs and fed her babies like a wild bird. So with rare birds as with human beings the rules are often broken as a function of individual experience.

CHAPTER TWELVE

Oedipus–Schmedipus

Jewish Mother 1: "What's the matter, you're crying? So unhappy yet!"

Jewish Mother 2: "It's my son. He's unhappy. His doctor says he's got an Oedipus complex."

Jewish Mother 1: "Oedipus—Schmedipus. What's the difference as long as he loves his mother?"

SIGMUND FREUD speculated that neurosis was the price human beings paid for living together in a tightly-integrated civilized society. In other words, as society expanded and the number and complexity of rules and regulations increased, civilization would require more suppressive control over each individual. More neurosis would be the result. We have watched Lucy to see if the development of her personality was consistent with this speculation of Freud's. It was not.

We have suppressed Lucy far more than chimpanzees in nature would be suppressed. The civilization in which Lucy lives requires that she not destroy furniture or plants; not play too rough; not bite; learn to use signs from the American Sign Language of the Deaf; not defecate or urinate except under prescribed circumstances, etc. Though she does not always conform to these pressures, as in her inconsistent response to toilet training, the pressures are there and she seems to experience them. In no way has she become neurotic. In my opinion she did not become neurotic even though she was subjected to more external control than chimpanzee nature is accustomed to, because she never experienced a deficit in social contact while growing up. Also, the controls were administered with affection and she could get away from them at times, for example, in the privacy of her own room. Man and chimpanzees are social beings; they require the presence of others of the same kind, particularly when growing up. (Our experience with Lucy suggests that to some extent the social needs of the chimp may be met by man, and vice versa.) I have seen neurotic and even psychotic chimpanzees in human homes, zoos, and in research colonies. These were chimpanzees who could not get along with other chimps or with human beings. They were constantly biting one another, or behaving in ways that got themselves rejected or bitten. Even when alone they did not seem tranquil; some pulled out their hair, pounded the walls in stereotyped fashion, or rocked repetitiously, repeatedly banging their heads against the wall like an autistic human child. Some could not master the social interaction necessary for copulation, though they tried their best. In all such cases the chimpanzee had been exposed to social isolation while growing up. That is, the neurotic or psychotic chimpanzee had been either raised in a cage alone for extended periods, often without even visual contact with another chimp; or if raised by humans, the chimpanzee was not allowed to have full contact with other humans or chimps. For example, I once knew a human mother who adopted a chimpanzee and maintained a symbiotic relationship with it. She kept it so close to her that she would not allow other people to play with it or hold it unless she controlled the interaction—the chimpanzee equivalent of the omnivorous, over-solicitous human mother who produces infantile, under-socialized children. By the time this chimp was three or four years old nobody but the mother could control him. He was uncomfortable and anxious around anyone else and bit people unpredictably. He eventually died when the mother was away on a trip, in my opinion from an acute, agitated depression which produced sufficient stress, so an infection which would normally have been of little consequence did him in. We therefore were very careful to maximize Lucy's contact with people, and we never punished her by "sending her to her room" or by depriving her of social contact in any form.

Not only are human beings and chimpanzees social animals, they both experience a long period of dependency upon the mother. Interpersonal dependency of this magnitude leaves both species peculiarly vulnerable to disruptions of the mothering process. Mothering, so critical at first that members of either species will die without it even if their physical needs are met, must gradually be reduced or the offspring does not develop the capacity to become even relatively independent. Lucy had social stimulation and contact exclusively from Jane for a year, then gradually from others both within and without the family. Gradually she became more independent—not exclusively so, for she still is a social animal and no chimpanzee is an island, either. However, Lucy now is independent enough that her social needs can be met by people other than Jane. Perhaps this is the meaning of true independence in humans as well; that we grow away from defining ourselves in terms of our parents and find sources of emotional fulfillment outside the family.

In Freud, the Oedipus complex was this pattern of attitudes: the child experiences sexual feelings for the parent of the opposite sex and fear and rivalry for the parent of the same sex. This family romance and painful triangle was thought to be an inevitable part of growing up and one reason for the universality of the incest taboo. That is, Freud thought that all human societies had a primary incest taboo because there was an instinctive wish to commit incest; there would be no reason to forbid an action unless there were wishes to commit it, he reasoned.

Lucy's behavior in this context is instructive. By virtue of her enormous dependency upon Jane she has developed protective maternal attitudes toward Jane; as mentioned, she will even threaten me if Jane indicates that I am hurting her. She watches Jane's every move and mood; she is happy when Jane is happy and distressed when Jane cries or is anxious. Yet when sexually aroused, in estrus, she will *not* make thrusting movements of her genitals on Jane's body, any more than she will on mine. Her built-in incest taboo extends to family members of both sexes. Yet I have seen her make thrusting movements of her enlarged genitals against the bodies of two other women. In a sense, her incest taboo may protect her from developing Oedipal problems by directing her sexuality outside the family.

When chimpanzees and man branched off from our common ancestor on the evolutionary tree, we humans got the short end of the stick. We inherited no biological basis for an incest taboo; no force other than social pressures to direct our sexuality outside the family. Though social pressures are sufficiently powerful that the act of incest is relatively rare, the wish probably is always there.

Indeed, what is more natural than the Oedipus complex? Our parents mold and shape our sexuality in many ways because they are the people we love the most, know best, and who love and care for us the most, (however ineffectually they may express it) at the time our sexual feelings are first experienced by us. It therefore should hardly be surprising that we have sexual feelings about parents. It is much more surprising that in the majority of cases overt sexuality is redirected toward non-family members, since humans probably do not have the biological help which simplifies Lucy's life. (I often felt grateful that I had Lucy rather than a human child for a daughter, because I could enjoy hugging and kissing her without fear that I was being too seductive for a father.)

It is a strange paradox that Sigmund Freud, the discoverer of the Oedipus complex and the first to recognize the importance of the universality of the incest taboo, should so often have committed and fostered psychological incest. In one myriad form or another Freud committed psychological incest, often with results as disastrous as my own experience with it. The first psychoanalysts trained by Freud who then became his

"She watches Jane's every move and mood; she is happy when Jane is happy and distressed when Jane cries or is anxious."

close friends, confidants, and colleagues often were tortured by suicides, psychotic breaks, neurotic depressions, and violent feuds after the psychologically incestuous relationship with him or between his "Sons and Daughters" erupted into convulsion. According to the psycho-historian Paul Roazen (*Brother Animal*, Knopf 1969), Freud even analyzed his daughter Anna, thus binding her to him for life. Though I feel hurt when Lucy goes into estrus and rejects me, it is not all rationalization when I tell myself, "It probably is for the best."

Lucy's biology protects her from incestuous relationships while we humans are dependent on social learning—and our vanity often stands in the way of learning. For much of the time that I was in an incestuous therapeutic relationship I "knew" with my head, but not with my heart, such painful consequences of incestuous relationships between therapists as these: Victor Tausk, one of Freud's most brilliant and creative pupils, committed suicide after being rejected by Freud. Because of his love and admiration for the master he could not express the pain and resentment the rejection caused him, except toward himself. Herbert Silberer, another pupil and close associate of Freud's, killed himself nine months after being totally rejected by Freud. Though it is considered a professional sin for a therapist to marry a patient, in the closely-knit family circle of early psychoanalysts it was common. Reich's first wife, Bernfeld's last wife, Rado's third wife and one of Fenichel's wives were either their patients or former patients (Roazen, 1969). To take one further example, the hostility between Freud and such "sons" as Adler and Jung, when they developed ideas of their own, is even better known. The development of autonomous selfhood depends upon breaking parental ties, and social or physical incest makes this more difficult, if not impossible.

". . . Girls are made of Sugar and Spice and Everything Nice"

In spite of her nearly symbiotic relationship with Jane, which would be a neurotically-dependent relationship were Lucy human, Lucy may have been spared neurosis for another reason. She avoided learning stereotyped sex-linked roles.

In the process of growing up, human beings all too often acquire sex-linked role stereotypes. For example, I early "learned" that boys were aggressive, tough and big boys don't cry. Girls, on the other hand, were passive, cooperative, tender, and tearful. When such roles become incorporated into the self-concept they function as a psychological straitjacket, reducing the spontaneity and flexibility of personality and behavior.

Because she was a chimpanzee, or perhaps because such concepts now seem artificial to Jane and me, we never imposed them upon Lucy. From the outset she played with toys without reference to whether they were "boys" or "girls" toys. She had dolls and plastic cars and soldiers; a tricycle; toy bulldozer; and crayons, coloring books, and finger paints. When she played "dress up," she wore Jane's or my clothes with equal frequency. She loved to put on my boots, shorts, or bathing suit and parade around the house, studying herself in the mirror. Just as often, whenever she could steal them or when Jane gave them to her as a treat, she wore Jane's sandals, blouses or skirts, and loved doing so. I could never discern a preference for toys or actions commonly labeled as "feminine" or "masculine." This may be one reason why Lucy seems so much freer to enjoy play than the average human being; this is not simply a species difference since many captive chimpanzees are not as playful as Lucy. In other words, Lucy may have no concept that "a nice girl" wouldn't do this or that, so her behavioral repertoire is not limited by sex-role stereotypes. I believe this lack of sex-role stereotyping has also made her sexual behavior more flexible.

In nature, wild female chimpanzees copulate by crouching and "presenting" their rears to the male. In estrus Lucy has never presented her genitals in this position, to my knowledge. She has, however, sexually stimulated herself by rubbing and thrusting her genitals against the bodies of both men and women (always excepting Jane, Steve or me) while sitting on their laps. Never to my knowledge has she assumed the sexual position of the wild female chimpanzee. This observation could be the result of Lucy's never having been in estrus around a male chimpanzee, or seen a female chimpanzee copulate in this position. In other words, I think it likely that her sexual behavior is flexible because she was not shaped by the chimpanzee equivalent of sex-role stereotyping in childhood.

CHAPTER THIRTEEN

Tomorrow

IN DECEMBER, 1974, Jane and I took our first trip together in almost ten years. Not since the day Lucy came home had Jane and I been away together—not for a single night. Even when Jane went to a hospital for major surgery I stayed with Lucy as both of us felt that Lucy's feelings of being alone were more likely to cause a dangerous depression than Jane's feeling of being alone while she had surgery. On three occasions I went to Europe to teach psychology courses, and once to Africa. Had we not had Lucy, Jane could have gone with me, but again we had felt it was too dangerous to leave her alone, so great were her needs for mothering. We had both seen infant chimpanzees die of marasmus, or of acute depressions, when adopted by human families and then treated as a "pet" rather than a member of the family and left alone when the parents took trips. We had not been willing to take chances with Lucy.

However, Lucy was full grown then and we doubted that she needed us so much. Steve had moved to town the year before and was going to the University, studying to be admitted to medical school. This book was almost completed and I wanted to go to San Francisco and discuss it with a prospective publisher. Jane, who was also going to school, had free time between semesters as did Steve. So Jane and I went to San Francisco leaving Lucy for a week with Steve and Nanuq.

We were gone for a week and we had a wonderful time together. We got to know each other again, and became closer than we had been at any time during the ten years with Lucy. We enjoyed one another thoroughly as we explored what to each of us is the loveliest city in America. On Christmas day we drove to Esalen at Big Sur, spending the day together at the baths, a place with much meaning for both of us. Returning to San Francisco, we met with the prospective publisher, Mrs. Peggy Granger of Science and Behavior Books and had an utterly delightful time with that charming and erudite woman. Peggy gave me the first professional editorial opinion that I had had since beginning to write this book which was as helpful as it was satisfying to receive. All in all it was one of those weeks, so rare in the average lifetime, when everything goes well. Not once did we worry about Steve or Lucy.

On the return Steve and Nanuq met us at the airport. He had left Lucy in her room though he had brought Nanuq.

On the way back from the airport Steve had told us about his week's experience. He had slept in my bedroom with Nanuq, leaving the door open while Lucy had slept down the hall in Jane's bedroom by herself, the door to which was left open. Everything had gone very well. In the mornings Lucy would wake them up, either by chasing Nanuq through the house, or by moving the furniture around in the living room so loudly that he would awaken. (It's okay for Lucy to move her furniture.) He told us that Lucy had begun to like Nanuq even more, to really love her and to demonstrate great affection, although their play sometimes became so rough that Nanuq, big and tough as she is, became frightened.

When we saw Lucy for the first time after our week's absence her greeting was different from what it had ever been before. She did not whoop, hoot, or scream excitedly from her roof-top room as we got out of the car. She simply watched us. When Steve opened her door she did not charge into the living room and jump into our arms, covering our mouths with hers as she had so frequently done when I had returned from trips. Instead, she came rather slowly into the room. She looked at both Jane and me, then walked around the room paying us little direct attention. I thought at first that she was mad at us for being gone and had decided to demonstrate her independence by ignoring us, but that was not the case at all. Instead she was greeting us more as one adult chimpanzee to another, for she then came to us and started to groom us, first Jane and then me. She stood on two legs in front of Jane making a lip-smacking noise while rapidly unbottoning Jane's blouse. After a moment or two of this she turned to me and groomed me in the same fashion, unbuttoning my shirt. After a few moments of the mutual grooming she went over and sat down on the couch and began to eat an apple, which she took from a bowl of fruit on the table. Nanuq sat beside her, Steve sat at the other end of the couch, Jane was on the opposite side of the room and I was sitting in my favorite chair. I looked at Lucy, who was calmly eating her apple while watching us, completely at home, completely relaxed, completely grown. Jane seemed more relaxed and alive than she had been in a long time. When I looked at Steve I realized he now was fully grown. He was leaving the next day for a back-packing trip between semesters. Suddenly I felt like breaking into tears. Here we were, our whole family; we had been through so much pain together and yet Steve and Lucy had made it; each had become so warmly human, such great people. In spite of everything our family had been a functional one. Steve could now take care of himself. He was an honor student, going to medical school in the fall. He loved us, as we loved him, but he no longer needed us. Lucy, too, was grown. It was clear that Jane and I could be gone without her becoming depressed, at least as long as she had human contact in our absence. We had raised a chimpanzee longer as a human child in isolation from her own species than any chimpanzee had ever previously been raised. Of the dozen or so chimpanzees people had tried to raise as humans, all but Lucy had died or had been returned to zoos or colonies when family life got too rough. We had explored the deepest depths of the psychobiological basis of being human, for here was Lucy, biologically a chimpanzee to be sure, but psychologically able to live healthily and happily as a human being given no more supervision than a mentally-retarded child would need. Yet we could not end here. We had to provide for Lucy. She has given us much knowledge and love, and we have exposed her to the joys and sorrows of human life for a longer time and more intensively than any chimpanzee has ever experienced them.

What about tomorrow? Lucy at ten will outlive both Jane and me, since she can be expected to live to at least 50 and perhaps 60 or 70 years of age, given continuous care.

Jane and I want to provide for Lucy as well as learn more from her and yet we want more freedom to live a normal life as well. So what can be done? Here are some alternatives.

A Chimpanzee Colony

Though Lucy taught us much and gave us great love and enriched our lives and our growth as people, it was a horrible thing to do to her. Though she is happy, and does not long for a jungle life she has never known, she was born beautifully adapted by evolution to the savanah and jungles of Tanzania. Living with us, however, she had no opportunity to learn chimpanzee social behavior—she would not know how to greet another chimp, when to search for food or share it, what jungle plants might be poisonous to her, how to satisfy her sexual needs, or how to protect herself. She will be scared to death when she sees a chimpanzee for the first time. To illustrate the cruelty it would take to simply lock her up in a zoo or colony I want to describe her typical day. I will select yesterday, since it is most fresh in my mind.

One Day in the Life of Lucy Temerlin

Lucy awakened at 7 a.m. after eight hours sleep with Jane in a king-sized bed on a Simmon's Beautyrest mattress. Still sleepy, she walked into the living room and sat on a Danish modern sofa while I fixed coffee for the three of us. Jane slept until the coffee was ready.

Perked-up by the coffee, Lucy made a circular nest of sofa cushions on the floor and sat in it for half an hour, looking at *Time, Newsweek,* and the *National Geographic.* We keep them on the coffee table because they seem to be her favorites, unless she is in estrus, then she prefers *Playgirl.* About the time she finished reading Jane had her breakfast ready—a bowl of oatmeal to which raisins and beef protein powder had been added, and a glass of grape Tang.

After breakfast she went to her room and played alone for about an hour and a half. Then she was visited by Roger Fouts and Barbara Mandell. They brought her into the living room where she sat on Barbara's lap most of the time while talking with Roger in ASL. That day Barbara and Roger were interested in how many times Lucy initiated conversations by signing as compared with her responding to their signing. After a half hour of language lessons they stopped for tea when Lucy went into the kitchen and started filling the teakettle with water. Lucy drank a cup of tea sweetened with honey.

When Roger and Barbara finished, they returned Lucy to her room. I had no therapy client at 12 that day, so I lunched with Lucy. She had two soft-boiled eggs, preferring four-minute eggs, a half pint of peach yogurt and an orange.

She spent the afternoon in her room, alternately sleeping and playing and watching the countryside. In the middle of the afternoon I took her a banana between appointments, and tickled her for two or three minutes.

Jane came home about 5:00 and I saw my last client at 6:00. Then the three of us had a gin and tonic together sitting in the living room. Then Lucy invited Jane and me to chase her. I didn't feel like it and was grateful to see Steve and Nanuq drive up. So Nanuq and Lucy played chase; they took turns chasing one another about the house while Steve talked with Jane and me. Steve decided he would not stay for dinner.

Lucy got hungry while Jane and I were having a second drink, went to the refrigerator and helped herself to a carton of raspberry yogurt, a few bites of left-over pot roast, a carrot, half a carton of partially-defrosted frozen strawberries, and took three or four bites out of a head of lettuce. Then she went back to the sofa, covered herself with her blanket, and fell asleep.

Put her in a chimpanzee colony? The thought brings to mind the Jewish intellectuals of Germany who were honored citizens of the most culturally and scientifically advanced nation in Europe one day, and found themselves without friends, property, or personhood the next, as they were herded behind the barbed wire of the concentration camp. Human-raised chimpanzees have been integrated into colonies of chimpanzees before, but it has never been done with a chimpanzee who has been with humans exclusively or for so long. Even with three or four-year-old chimps, who would be more flexible than Lucy, it has to be a gradual process, and the chimpanzee experiences much fear, terror and pain before becoming a chimp among chimps. Often chimpanzee madness results, even with the best of gradual initiation rites. We have no contact with a colony that would allow us to gradually introduce her, maintain contact with her, and guarantee that she not be deprived of social contact with the chimps and people she likes, and that she will not be used as a subject in medical research.

Zoos

A zoo would not work for the same reasons. That is, zoos frequently maintain them in an uninteresting, sterile environment. A zoo might have five to twenty chimps in an area the size Lucy is accustomed to, or what is worse, they often keep them in cages alone.

Research

Because of Lucy's unique status she is a unique subject for research. Jane and I are, of course, far more interested in our daughter's welfare than in contributing to science, but here are a few examples of research possibilities which would simultaneously contribute to science and to Lucy's welfare.

1. Research on mothering. If Lucy were artificially inseminated with semen obtained from colony chimpanzees and then became pregnant, a study of her maternal behavior would answer such questions as whether or not chimpanzee mothering behavior has to be learned. My prediction would be that Lucy would mother her baby, as Jane mothered her, but we do not know.

This experiment would allow her to actualize her potential for motherhood, if there is such a thing, but would contribute nothing to her capacity to get along with other chimpanzees.

2. We could try and breed her to an adult male chimpanzee. Lucy would be scared to death. She would have to be introduced to the stud very carefully and gradually. Furthermore, Lucy's mate would have to be selected with the utmost care.

Indeed, I have strong feelings about the qualifications which must be possessed by the future husband of my darling virginal daughter. He would not have to be Jewish, but he would have to be gentle and patient; and he should have enough tolerance and passivity to let her take the initiative. Otherwise, I am sure, Lucy would be terrified into chimpanzee madness. Ideally the male would be tame enough for us to handle so that we could easily take him to and from brief meetings with Lucy until she learned to get along with him.

3. Since species-isolated chimps are terrified when introduced to other chimps for the first time, even when they have been isolated for much less time than Lucy has, we might give her a baby chimp. If we introduced her to a three-year-old chimp, for example, since Lucy would be so much larger it might reduce her threat and they might then become playmates. If we do this, it would be fascinating to see if Lucy tries to communicate with the newcomer using the sounds and gestures of wild chimpanzees, the ASL she has learned, or both.

These are possibilities we are considering at this time.

I was raised in the romantic tradition and I like books to have happy endings. If they do not have happy endings they should have tragic endings. I hate books which have no ending—like this one. The story of Lucy Temerlin is not finished. Jane and Steve and I talk constantly about it, but it is very complicated, for we want to live normal lives now, though we are still committed to Lucy. All I can say definitely at the moment is that part of the earnings from this book will be used to establish a trust fund for Lucy, to provide for her care and comfort throughout her life, and that Jane or I or both of us will write another book, so that you will finally have an ending. If you have any ideas, write to me at Science and Behavior Books, P.O. Box 11457, Palo Alto, California 94306.

PUBLISHER'S NOTE

Science and Behavior Books has contracted for a second book from Maury and one from Jane, so Lucy's story will be a continuing one.

BIBLIOGRAPHY

SUGGESTED READINGS

The classical studies of chimpanzees in the wild by Dr. Goodall as a participant observer are as follows.

Lawick-Goodall, Jane van. *My Friends the Wild Chimpanzees.* Washington: National Geographic Society, 1967.

Lawick-Goodall, Jane van. *In the Shadow of Man.* Boston: Haughton Mifflin Company, 1971.

Dr. Goodall's work in the Gombe Stream area of Tanzania is presented in more detailed and technical form in Vol. 1, Part 3 (1968), pp. 161-311 of *Animal Behavior Monographs.*

CHIMPANZEES IN CAPTIVITY

Hahn, Emily. *On the Side of the Apes.* New York: Thomas Crowell Co., 1972.

Hayes, Catherine. *The Ape in our House.* New York: Harper & Row, 1 51.

Kellogg, W. N. and Kellogg, L. A. *The Ape and the Child.* New York, McGraw Hill, 1933.

Savage, E.S., Temerlin, J. W., and Lemmon, W. B. "The Appearance of Mothering Behavior Toward a Kitten by a Human-Reared Chimpanzee." Proceedings of the 5th Congress of the International Primatological Society. 1974 in Press.

Yerkes, Robert M. "Chimpanzees: A Laboratory Colony." New Haven: Yale University Press, 1943.

ON TEACHING CHIMPANZEES ASL

Fouts, Roger S. *Man-Chimpanzee Communication.* I. Eibl-Eibesfeldt, G. Kurth (eds.) Hominization of Behavior. Gustav-Fisher and Verlog-Stuttgart. In Press.

Fouts, Roger S. *Talking with Chimpanzees.* Science Year, 1974, pp. 34-49, Field Enterprises, Chicago, 1974.

Gardner, Allen and Beatrice. *Teaching Sign Language to a Chimpanzee,* Science, Vol. 165 (1969) pp. 664-72.

Gardner, B. T. & Gardner, R. A. *Two-way Communication with an Infant Chimpanzee.* Schrier, A. M. and Stollnitz, F. eds. Behavior of Non-human Primates. Vol. 4, 117-184, Academic Press. New York, 1971.

General books on animal behavior and non-human primates other than chimpanzees.

Altmann, Stuart A. ed. "Social Communication among Primates." Chicago: University of Chicago Press, 1967.

DeVore, Irven, ed. "Primate Behavior." New York: Holt, Rinehart and Winston, 1965.

Jay, Phyllis, ed. "Primates: Studies in Adaptation and Variability." New York: Holt, Rinehart and Winston, 1968.

Lorenz, Konrad. *King Solomons Ring.* New York: Crowell, 1952.

Lorenz, Konrad. *On Aggression.* New York: Bantam Books, 1971.

Morris, Desmond, ed. "Primate Ethnology." Chicago: Aldine Publishing Company, 1967.

Washburn, S. L. & Jay, P. C. eds. "Perspectives on Human Evolution." Holt, Rinehart & Winston, Inc., 1968.

The following books and articles on psychotherapy and personality change do justice to the complexities and paradoxes of the process, including the role of the personality and inner experience of the psychotherapist.

Fagan, Joen & Shepherd, Irma Lee. *Gestalt Therapy Now.* Palo Alto, Science & Behavior Books, 1970.

Kopp, Sheldon B. *If You Meet the Buddha on the Road, Kill Him!* Palo Alto: Science and Behavior Books, 1972.

Kopp, Sheldon B. *The Hanged Man.* Palo Alto, Science and Behavior Books, 1974.

Kopp, Sheldon B. *Guru.* Palo Alto, Science and Behavior Books, 1971.

Lidz, Theodore. *The Person.* New York: Basic Books, 1968.

Perls, Fritz. *The Gestalt Approach and Eye Witness to Therapy.* Palo Alto, Science and Behavior Books, 1973.

Roazen, Paul. *Brother Animal.* New York: Knopf, 1969.

For readers interested in what Maslow called peak experiences and writers in earlier traditions referred to as "oceanic" or "transcendential" experiences, the following are good source books.

James, William. *The Varieties of Religious Experience.* New Jersey: University Books, 1963.

Tart, Charles T. ed. *Altered States of Consciousness.* New York: John Wiley and Sons, Inc. 1969.

White, John. ed. *The Highest State of Consciousness.* New York: Doubleday & Company, 1972.

The articles referred to in the chapter, "Keeping it in the Family," were:

Baumgold, John. Temerlin, Maurice K. and Ragland, Robert. Experience of Freedom to Choose in Mental Health, Neurosis and Psychosis. *Psychological Reports,* 1965, 16, pp. 957-962.

Gatch, Vera and Temerlin, Maurice K. "The effect of the belief in psychic determinism on the behavior of the psychotherapist." *Review of Existential Psychology and Psychiatry,* Vol. 1, #1, Winter, 1965.

Klein, Helen and Temerlin, Maurice K. "On Expert Testimony in Sanity Hearings." *Journal of Nervous and Mental Disease.* Vol. 49, #5, 1969.

Lee, Steve and Temerlin, Maurice K. "Social Class, Diagnosis, and Prognosis for Psychotherapy. *Psychotherapy: Theory, Research & Practice.* Vol. 7, No. 3, Fall, 1970.

Temerlin, Maurice K. "On choice and responsibility in a humanistic psychotherapy. *Journal of Humanistic Psychology,* Spring, 1963. Also in *Humanistic Viewpoints in Psychology,* F. T. Sevrin, Editor, McGraw-Hill. 1965.

Temerlin, Maurice K. and Trousdale, W. W. "The social psychology of clinical diagnosis." *Psychotherapy: Theory, Research, and Practice.* Vol. 6, No. 1, Winter, 1969.

Temerlin, Maurice K. "Suggestion effects in Psychiatric Diagnosis." *Journal of Nervous and Mental Diseases.* Vol. 147, No. 4, 1968. Also in *The Making of a Mental Patient,* R. Price and B. Denner, editors, New York: Holt, Rinehart, Winston 1973 and included in Scheff, Thomas J. Editor, *Labeling Madness,* Prentice Hall, Inc. 1975.

Temerlin, Maurice K. "Diagnostic Bias in Community Mental Health." *Journal of Community Mental Health,* Vol. 6, #2, 1970. Also in *Behavior Disorders: Perspectives and Trends,* third edition, O. Milton and R. Wahler, Editors, Philadelphia: J. B. Lippincott, 1973.

On the development of sexuality in the female, these books are excellent.

Bardwick, Judith M. *Psychology of Women: A Study of Bio-Cultural Conflicts.* New York: Harper and Row, 1971.

Sherfey, Mary Jane. *The Nature and Evolution of Female Sexuality.* New York: Random House 1966.